T0344466

Political Economy of Housing in Chile

Through the lens of political economy, this book positions housing as a key factor in understanding social inequality. It does so by drawing on rich empirical evidence from the case of the Chilean housing market.

This book provides insights on the articulation between real estate development, housing provision and social inequality based on applied urban economics analyses that illustrate the contradictions of neoliberal urbanism through the case of Chile. For neoliberal urbanism, the good city is not equal for all, it is based on the principle of profitability and benefits from segregation to make capital investment more efficient. The chapters of this book expose how these processes are generated by a political system that allows them rather than by the invisible hand of the market.

The book will be of interest to graduate students in urban studies, urban planning, sociology and urban geography. It will also appeal to decision-makers and also to actors in the real estate market seeking to perfect the social benefits of their professional activities, aspiring to generate more egalitarian and just cities.

Francisco Vergara-Perucich is an architect from Universidad Central de Chile, Master in Architecture from Pontificia Universidad Católica de Chile, and MSc Building and Urban Design in Development and Doctor in Development Planning from The Bartlett Development Planning Unit, University College London. He currently directs the Centro Producción del Espacio at Universidad de las Américas.

Carlos Aguirre-Nuñez is a civil constructor from Pontificia Universidad Católica de Chile and Master and PhD in Urban Management and Valuation from Universidad Politécnica de Catalunya. He is currently Researcher of Urban Studies at Universidad San Sebastián in Chile.

Felipe Encinas is an architect from Pontificia Universidad Católica de Chile, MSc from the University of Nottingham in the UK, and PhD from the Catholic University of Louvain in Belgium. He is currently Associate Professor at the School of Architecture at Pontificia Universidad Católica de Chile and a researcher at the Centre for Sustainable Urban Development (CEDEUS).

Rodrigo Hidalgo-Dattwyler is a geographer from Pontificia Universidad Católica de Chile and PhD in Human Geography from the University of Barcelona. In 2014 he received the National Geography Prize awarded by the Chilean Society of Geographical Sciences of Chile. He is currently head of the PhD Programme in Geography at Pontificia Universidad Católica de Chile and Director of the Revista de Geografía Norte Grande.

Ricardo Truffello is Geographer and Master in Geography and Geomatics from the Pontificia Universidad Católica de Chile and PhD in Engineering in Complex Systems from the Universidad Adolfo Ibáñez de Chile. He currently directs the Observatorio de Ciudades at the Pontificia Universidad Católica de Chile.

Felipe Ladrón de Guevara is an architect and holds a master's degree in Urban Planning from the Pontificia Universidad Católica de Chile. He is currently Assistant Professor at the School of Architecture and Head of the Analysis and Quality Assurance Unit of the Faculty of Architecture, Design and Urban Studies of Pontificia Universidad Católica de Chile.

Routledge Studies in Urbanism and the City

Public Participation Process in Urban Planning
Evaluation Approaches of Fairness and Effectiveness Criteria of Planning Advisory Committees
By Kamal Uddin, Bhuiyan Monwar Alam

Hegemony, Security Infrastructures and the Politics of Crime
Everyday Experiences in South Africa
Gideon van Riet

Marketplaces
Movements, Representations and Practices
Edited by Ceren Sezer and Rianne van Melik

Designing Healthy and Liveable Cities
Creating Sustainable Urban Regeneration
Marichela Sepe

The Political Economy of Land
Rent, Financialization and Resistance
Edited by Mika Hyötyläinen and Robert Beauregard

Future of the City Centre
Global Perspectives
Edited by Bob Giddings and Robert J Rogerson

Political Economy of Housing in Chile
Francisco Vergara-Perucich, Carlos Aguirre-Nuñez, Felipe Encinas, Rodrigo Hidalgo-Dattwyler, Ricardo Truffello and Felipe Ladrón de Guevara

For more information about this series, please visit www.routledge.com/Routledge-Studies-in-Urbanism-and-the-City/book-series/RSUC

Political Economy of Housing in Chile

Francisco Vergara-Perucich, Carlos Aguirre-Nuñez, Felipe Encinas, Rodrigo Hidalgo-Dattwyler, Ricardo Truffello and Felipe Ladrón de Guevara

LONDON AND NEW YORK

First published 2023
by Routledge
4 Park Square, Milton Park, Abingdon, Oxon OX14 4RN

and by Routledge
605 Third Avenue, New York, NY 10158

Routledge is an imprint of the Taylor & Francis Group, an informa business

British Library Cataloguing-in-Publication Data
A catalogue record for this book is available from the British Library

Library of Congress Cataloging-in-Publication Data
Names: Vergara Perucich, Francisco, author. | Aguirre-Nuñez, Carlos, author. | Encinas Pino, Felipe, author.
Title: Political economy of housing in Chile / Francisco Vergara-Perucich, Carlos Aguirre-Nuñez, Felipe Encinas, Rodrigo Hidalgo-Dattwyler, Ricardo Truffello and Felipe Ladrón de Guevara.
Description: First Edition. | New York, NY : Routledge, 2022. | Series: Routledge studies in urbanism and the city | Includes bibliographical references and index.
Identifiers: LCCN 2022043803 (print) | LCCN 2022043804 (ebook) | ISBN 9781032391793 (hardback) | ISBN 9781032391823 (paperback) | ISBN 9781003348771 (ebook)
Subjects: LCSH: Housing—Chile. | Housing policy—Chile. | City planning—Chile. | Capital investments—Chile. | Equality—Chile.
Classification: LCC HD7324.A3 V47 2022 (print) | LCC HD7324. A3 (ebook) | DDC 363.5/5610983—dc23/eng/20221031
LC record available at https://lccn.loc.gov/2022043803
LC ebook record available at https://lccn.loc.gov/2022043804

ISBN: 978-1-032-39179-3 (hbk)
ISBN: 978-1-032-39182-3 (pbk)
ISBN: 978-1-003-34877-1 (ebk)

DOI: 10.4324/9781003348771

Typeset in Times New Roman
by Apex CoVantage, LLC

Contents

Figures

Tables

Preface

This book articulates the problem of the housing crisis with the political economy in order to advance the complexity of a social problem of great global relevance, broadening its purely operational scope, which refers to the material production of housing. The aim of the book is to discuss the scope of housing in terms of the integral planning of the state, the nation and economic relations. Housing is a result of political action, and as such, its scope goes far beyond architects, planners or urban planners to incorporate actors as diverse as investors in stock exchanges to presidents of a central bank. If the housing problem has not found lasting solutions in much of the world, it may be because of its prioritisation, as a problem of public interest is mainly sectoral rather than a comprehensive matter affecting the many. In our view, the housing problem cannot be solved only with design, politics or economics, but with all of them at the same time, interrelated, and by incorporating the knowledge of other disciplinary fields ranging from anthropology to social work. If space is a result of social relations, housing is possibly the most social of all spaces.

The book we present is the result of years of joint work by a group of authors in different academic instances. Our reflections are situated in Chile, where we have been part of a real struggle to broaden the decision-making frameworks that affect the process of producing decent housing for the population. In our work as researchers, we often come across findings that deserve to be heard by decision-makers. However, neither the findings nor the researchers themselves have the same information dissemination apparatus as the actors who have vested interests in the field. A scientific article does not carry the same weight as several press releases in the eyes of decision-makers, and often, having an open platform in the media allows scientific findings to be disqualified without counterweight. On the other hand, researchers tend to fall into the innocence of believing that findings alone, backed by science, should have enough weight of their own to impute the arguments of an industry that defends its business. This permissive innocence often allows the findings to fail to influence housing policy, which is our focus of study. But we have learned from both failures and achievements. One of the main lessons learned is that the housing problem is a case of political economy, according to a study on inequality that arises from productive and commercial relations with the state and society. This book seeks, precisely, to share the process that led to this learning.

The authors have gathered findings that allow us to support that the housing problem is a matter of political economy. We have reported how energy poverty is reflected in housing design and production processes, emphasising that the real estate industry needs to adapt its ways of producing for the urgencies of climate change. We have also raised the importance of improving the mechanisms for measuring socio-territorial realities, so that the authorities understand the future value of this data in designing public policies. We have also produced evidence that indicates the contradictions of the housing market and how its results are unattainable for many households that do not have secure access to housing. Framing these discussions in a political economy book allows us to broadly share how our research findings align with a broader discussion of structural inequality in neoliberalised nations.

In 2020, we launched a book in Chile that served as a general basis for this book. At that time, the discussion we developed was much more focused on the local problem. With the pandemic years of the new coronavirus and its consequent social crisis, we saw how our findings could be useful in other nations, where the ways of understanding the housing problem could be aligned with the issues that arise in Chile. To this, we added that Chile is a global outlier, being the first nation where a neoliberal political economy model was imposed in 1975 under the dictatorial regime of Augusto Pinochet, advised by the Chicago Boys and Milton Friedman himself. Chile was the guinea pig for the neoliberal experiment that was later implemented in a lot of places around the world. Therefore, as an atypical case of neoliberalism, almost caricatured, the lessons that emerge from its contradictions can be useful in other nations where neoliberalism is also in crisis. In this book, we share applicable and replicable methodological approaches with data that can be used in other countries to compare experiences and review the scope of the political economy of housing in each specific local reality. In addition, in the 2020 book, we knew little of the transcendental effects that the social explosion in Chile would have, and this book did not know the importance the new coronavirus would have in revealing the importance of housing but from the disciplines of public health, thereby broadening the scope of the problem that at that time seemed very sectoral, especially in terms of urban planning disciplines.

We have tried to develop a didactic book, avoiding technicalities and focusing on how the results speak to the problem of inequality from the analysis of housing. In essence, this is a book that uses housing to explain one of the possible origins of inequality: the fair distribution of territory and the absence of the right to the city. As we write this book, Chile is debating the approval of a new Political Constitution of the Republic, drafted for the first time in history by a democratically elected, parity-based and socially diverse collegiate body. This new constitutional text enshrines the right to housing and the right to the city. If this constitution is implemented, this book could serve to identify factors in the housing production process that facilitate the reproduction of inequality in order to generate solutions from the new rights-based approach.

1 Introduction

> The production and reproduction of space are profoundly ideological/political events. Space is not merely an inert substance deployed by designers. Overall, urban planning adjudicates the process as the agent of state interventionism, on behalf of social reproduction, capital accumulation and the circulation of commodities. Therefore, it is centrally involved in the management of a key factor of production, namely land (and, as such, in the generation of land rent, profit and surplus value).
>
> Alexander Cuthbert, 2011[1]

The process of urban development is profoundly ideological. Dressing up the disciplines dedicated to the city in scientific suits, with an aspiration to develop results in an apolitical, ideology-free way, is a mistake of origin. It fails to recognise that space, and especially the city, is produced politically. Disputes, impositions, violence, community organisation, resistance and agreements shape-built space. Under this premise, in our understanding, there is no technocracy without ideologies, given that technique is always inscribed in epistemological frameworks that generate interpretations marked by the authors of each design, originating in specific ontological positions. In other words, subjectivity is also part of the design of the methods used in urban studies. This may be a truism in some parts of the world, where science is indeed closer to decision-makers, but we who write from South America know that it is a necessary distinction. So, urbanism, like many other disciplinary fields, allows us to justify approaches to urban productive processes in search of generating convictions and certainties about how to operate in the city, but it serves specific purposes and, therefore, has a strong ideological component at its root. In Chile, the basic ideology of urban planning since its neoliberalisation began in 1976 (with the liberalisation of taxes on construction production processes) is to seek mechanisms that validate decision-making based on the profitability (economic and

DOI: 10.4324/9781003348771-1

social) of urban transformations. However, this is not exclusive to Chile, although it is one of the few places in the world where the phenomenon has been widely studied over 40 years and in the case of Santiago, it is more visible and therefore more interesting to analyse. Studying the processes of housing production in Chile is not so different from studying its socio-economic history. Peggy Deamer puts it clearly: Since the production of architecture and cities requires immense amounts of capital that articulate the public, the private and the social, the history of cities is in large part the history of capitalism itself.[2] Similarly, we will understand the neoliberal model in the sense proposed by David Harvey: Incubated during the crisis of the 1970s, masked under a "thick layer of rhetoric about individual freedom, personal responsibility, the virtues of privatisation, free markets and free trade", and which in practice legitimised "draconian policies aimed at restoring and consolidating the power of the capitalist class".[3]

It is for this reason that the book we present here is so relevant for the interpretation of history in Chile, given that it uses data and records from 2019, a key year in setting a different course from the turbo-capitalism to which we have been exposed since 1975. It is an important date, and although we do not know the future yet, the data indicate that many things will change from now on, including the political economy of housing and the disciplinary ethics that shape it. Referring to the national contingency, Sebastian Edwards, a renowned academic at the University of California (UCLA), said, "the neoliberal experiment is completely dead".[4]

In the early hours of Friday 15 November 2019, Congress signed the agreement to initiate a new constituent process to be plebiscite in April 2020. This book was completed on the same Friday. In seeking to make the ideology of this text transparent, we establish that we are critical of the processes that generate socio-spatial inequality and seek to understand how these end up benefiting a few at the expense of many others. Thus, we are inclined towards an urban ontology influenced by spatial justice thinking derived from the critical urbanism advocated by Jane Jacobs, Susan Fainstein, David Harvey, Raquel Rolnik and mostly, the principles of Henri Lefebvre's Right to the City. However, we also believe that contradictions underlie the very methodologies of neoliberal urbanism that attempt to justify its decision-making. It is for this reason that, from critical urbanism as an interpretative framework, we use the tools of neoliberal urbanism to study its conditions and review, with its own methods of validation, the fissures that emerge and materialise in conflictive socio-spatial phenomena. Above all, as a major ethical framework, we try to situate ourselves within the charter of human rights as a referential platform for our reflections. We believe that housing needs special attention from this perspective in the Chilean context.

Housing is a human right. It sounds like a slogan, but there is no chance of a good life without access to safe, healthy, well-designed and good quality housing. Homelessness is the lack of a bed, a bathroom, a kitchen, access to drinking water, shelter, privacy—the material certainty of the old saying "having somewhere to drop dead". The horrifying thing is that currently in Chile, without money, there is nowhere to drop dead. Moreover, legislatively, it is a human right because it is suggested by the United Nations, the planetary body in charge of safeguarding respect for human rights. In this respect, Leilani Farha, special rapporteur on housing for this organisation, after learning about the Chilean reality in terms of the right to housing, recommended that the State of Chile "redouble its efforts to diversify the housing model, ensuring that the creation of housing does not depend solely on private developers and ensure that families are not obliged to obtain a mortgage".[5] This accurate recommendation highlights one of the main problems facing the nation in ensuring the human right to housing: the privatisation of its allocation mechanisms and financialisation.

Housing in Chile is a private property rather than a right, and as such, it is traded commercially. In the current conditions of financialisation of everyday life, housing, as private property, has become a long-term fixed income financial asset. In other words, housing is used to invest and generate capital as its value increases over time. This financial function of housing is displacing its use function—today it is more important that housing is profitable than it is of good quality. An example of this is how the elemental proposal was interpreted to privilege the location of social housing and thus maximise resources to obtain better land instead of better housing.[6] While the architectural world paid a lot of attention to the design driven by the versatile skills of the elemental team, the decision-makers valued more that the design be in more central urban sectors to improve added value. In other words, they did not care as much about liveability as they did about its value increasing over time. This discourse permeated into other modes of urban production, and nowadays, the construction of 17 m^2 flats—known as nanoapartments—is beginning to be validated as a possibility to live close to the centre and, therefore, with good added value. This is aberrant, but it follows the neoliberal logic of *Homo economicus*, in which human decisions are assumed to be based on economic views of problems and the decisions are rational. Unfortunately, while the process of acquiring a home can indeed be approached from an economic perspective, everyday life does not fit into such calculations. Its fundamental social role means that housing is not a matter for architects interested in its functionality or economists interested in its profitability, but is a problem that concerns all social strata, including health—ensured by an adequate design to avoid respiratory diseases—so that alienation processes do not occur due to models of housing that are not very inclusive to its technical, sustainable, legislative and political qualities, among many other aspects in which housing is key. However, in the national context—which is far from idealism—we have to analyse housing as an economic element, understand its complexity, and with it, deconstruct it to see if its productive processes and uses can be humanised. This is why we speak of the political economy of housing.

For David Ricardo, one of the fathers of political economy, economic relations are in themselves decisional and therefore, they fulfil a political function. This is why it is so relevant when it comes to understanding productive processes as well as such material aspects such as the formation of prices, rents, and values. For his part, Karl Marx states that value and price are formed as a consequence of a set of social relations that develop according to what the productive structures of society allow, which are created and formed according to the interests of the capitalist or, in this case, of the actors in the productive cycle with the greatest decision-making power. In Chile, it would be equivalent to what Gabriel Salazar has called the "civil political class" and the "speculative mercantile class".[7] So, to define the intellectual position from which this work is developed, we will understand political economy as the study of production and exchange in relation to legislative aspects, market, government, state and consumers (once citizens) from a critical perspective of income distribution, political-economic opportunities and aspects related to wealth. The political economy of housing is the study of the factors that explain its accessibility or inaccessibility, seeking to ensure its correct allocation to households from the market, the state and the tools provided by public policy. In the particular case of Chile, this study of the political economy of housing stems from the crisis of access that the country is experiencing. Although it is an issue that has been around for some time in academic circles, it gained particular notoriety a few months ago when it was mentioned publicly by the Chilean Chamber of Construction (Cámara Chilena de la Construcción). Indeed, in a meeting before authorities and businessmen—including the President of the Republic, Sebastián Piñera—the president of the trade union organisation, Patricio Donoso, declared, "I am not exaggerating if I say that we are facing a crisis of access to housing that is soon to turn into a social crisis".[8] And then, the social outburst that began on 18 October 2019 only confirmed it, since one of its declared causes was the high cost of living, where housing is the heaviest element.

The book we present here compiles a set of in-depth studies on the political economy of housing, trying to explain why it has become inaccessible and testing some ways to reverse this condition. Part of these chapters and their conclusions have already been presented in articles in specialised journals, book chapters and articles published in the prestigious Chilean media called *CIPER* (Centro de Investigación Periodística). This academic production has been developed within the Centro de Producción del Espacio of the Universidad de Las Américas and the Centro de Desarrollo Urbano Sustentable (CEDEUS) of the Pontificia Universidad Católica de Chile. Thus, the studies presented in this text seek to dispute with scientifically developed data that the housing market in Chile is imperfect and requires substantial improvements in order to advance towards the human right to housing.

The first chapter of the book discusses the context of housing in Chile, marked by its Political Constitution, where housing as such is not present in the text and it is treated only as private property. The second part explores the relationship between the cost of living in Chile's cities and household income—which is largely the justification for this study—given that housing is the most significant household expense and has now become unaffordable for large portions of the population. A third part sets out the arguments most commonly used to explain high housing prices by academics and the business community. The fourth part puts forward one of our hypotheses based on concrete data: housing is unaffordable because it has become a business that aims at profitability rather than habitability. The fifth part explores this idea further and discusses how housing has become a long-term fixed income financial asset, more profitable than other financial instruments and tremendously competitive for investors, thus breaking its use value and more strongly embracing its exchange value as the main criterion when deciding where to invest in housing and how to design it. In the sixth part, we extend a hypothesis on the importance of location as a key factor in defining housing prices and how this impacts social segregation. Towards the end, we propose a series of measures based on the necessary recognition of the social role of urban land and the importance of having an active state system of housing provision, along with strict regulatory frameworks for private residence, ensuring its fair price and high quality. This last part is based on international cases that could be applied in Chile. Finally, we develop a set of reflections and conclusions that seek to feed the upcoming debate on a new Constitution and the opportunities that open up from this space to rethink our cities.

Notes

1 Cuthbert, Alexander R Cuthbert, *Understanding Cities: Method in Urban Design* (London: Routledge, 2011), 78.
2 Deamer, Peggy Deamer, *Architecture and capitalism: 1845 to the present*, 2013, https://doi.org/10.4324/9780203499023.
3 Harvey, David. *The Enigma of Capital and the Crises of Capitalism* (Madrid: Ediciones Akal S. A, 2012).

4 Edwards, Sebastián. The Reality of Inequality and Its Perception: Chile's Paradox Explained. *Pro-Market*, 19 November 2019. https://promarket.org/the-reality-of-inequality-and-its-perception-chiles-paradox-explained/
5 Farha, Leilani. 2017. *Report of the Special Rapporteur on Adequate Housing as a Component of the Right to an Adequate Standard of Living, and on the Right to Nondiscrimination in this Context, on her Mission to Chile*. New York: United Nations.
6 A reflection on this extra-disciplinary understanding is developed by journalist Hector Cossio in *El Mostrador* (http://shorturl.at/vCUW1) and the Hyatt Foundation itself in the statement of the arguments for awarding Aravena the Pritzker Prize: So-called incremental housing allows social housing to be built on land that is more expensive but closer to economic opportunities and gives residents a sense of personal investment. At: www.pritzkerprize.com/announcement-ale-jan-dro-ara-ve-na.
7 Salazar, Gabriel Salazar, *La enervante levedad historica de la clase politica civil (Chile 1900–1973)* (Santiago: Debate, 2015); Gabriel Salazar, *Historia de la acumulación capitalista en Chile (apuntes de clases)* (Santiago: LOM Ediciones, 2003).
8 *Pulse*. Construcción advierte riesgo de crisis social por creciente costo de las viviendas. *La Tercera Pulso*, 10 May 2019.

2 Constitution and housing

This chapter reflects on one of the fundamental aspects of the political economy of housing: its orientation from the law in the Political Constitution of the Republic of Chile. In the last four decades, the current constitutional document, known as the 1980 Constitution, despite having undergone numerous reforms in the period following the military dictatorship of General Augusto Pinochet, has been vital to ensure the fulfilment of the political-economic project founded on the ideology of the so-called *Chicago Boys*,[1] a group of Chilean economists who specialised at the University of Chicago under Milton Friedman and later had the opportunity to implement these ideas in Chile during the dictatorship, acting as ministers, undersecretaries and heads of government services. Their project was based on the logic of maximising the economic efficiency of large groups of owners, where the state in a subsidiary way generates a set of rules that tend to promote a monetarist economic model,[2] known today as the neoliberal model.

Housing was not directly incorporated in the 1980 Constitution or in its subsequent reforms. This omission can be understood from the observation that housing is simply considered another consumer good that does not merit special treatment. It will be the market—through the rule of supply and demand—that will ensure the correct placement of houses in cities. From this perspective, only in cases where the market is unable to ensure the correct allocation of houses (or any other consumer good) should the state exercise its subsidiary role and contribute to filling the gaps in the availability of goods or resources with public initiatives. In other words, the subsidiary logic will use fiscal funds to subsidise the lack of market solutions only in those cases where parts of the population are excluded from access to essential goods and services. From a practical reading of the Constitution, it is explained that in the long run, the absence of housing as an explicitly declared right and as a counterpart to the strong private property regime that our nation's fundamental charter promotes ends up

DOI: 10.4324/9781003348771-2

producing a confusion between the idea of home and property, and the idea of being and having, thus opening the door to using housing as an asset to generate income rather than as a basic service. It is not surprising then that housing in Chile is used for economic speculation and exposed to risky financialisation.

Given that space (housing, city, territories, air) is absent in the Political Constitution of the Republic of Chile, we could say that this document is not complete. This is worrying, as it could be at the root of why Chile has such segregated cities.[3] Given the importance of the fundamental charter in determining the country's designs in so many aspects and considering that space is a social production,[4] the policies related to space that are not represented in this document make it difficult for housing to be placed at the centre of public discussion, as it is a problem that is often made invisible, and with it, its socio-spatial problems. This book was written precisely when the housing question in Chile was beginning to be less invisible and citizens were beginning to understand that it is not an individual problem, but a social and, therefore, a collective one.

At the same time, space appears subterraneously from other clauses that do not fully reveal its importance for society. In Article 3, the Constitution states that Chile is a unitary territory, which is another way of saying that its spatiality unites nationals in a set of political-administrative limits. In this sense, it does not mention borders as complex spaces of high social, cultural, economic and political richness, thus ignoring the importance of informing how the spatial form of the national territory is configured. The Constitution also fails to recognise the importance of urban space, problematise the city, account for its political role and establish a right of access to it. It also does not discuss housing or the house—the latter in the sense of "home" proposed by Bourdieu[5]—nor does it present them as fundamental parts of citizenship in Chile. In the Constitution, the only time the house or home is mentioned is to refer to the fact that its limits cannot be violated, that is, it is established that the state cannot violate private space. Once again, the house or home is defended as private property, a place outside the public sphere—in other words, a space alienated from the commons.

The Constitution of the Republic of Chile states that "no one may, in any case, be deprived of his property".[6] Not even expropriation for public purposes is immune from the right to private property, given that it is established that "the expropriated party may claim the legality of the expropriation act before the ordinary courts and will always have the right to compensation for the property damage effectively caused".[7] The social space is not relevant, but the ownership of the space is. The space can be destroyed, exploited, transferred, altered and, in general, whatever the owner wishes to do with it, be it a person, a company or the State of Chile itself. In synthesis, space is a social product[8] that has been financialised,[9] neoliberalised,[10] verticalised,[11] gentrified,[12] dehumanised[13] and exposed to precarious conditions.[14] Space needs to fix its legal body, its ideological constructions and mechanisms to resolve its contradictions, to advance justice and reconciliation between space and society.

But even so, without recognising housing directly, the Political Constitution of the Republic of Chile, in Article 4, establishes that Chile is a democratic republic and it is in the understanding of democracy that housing could be challenged in the current Constitution as equality before the law and, therefore, contrary to inequalities even when they are based on deficiencies in the economic model or the imperfection of the markets. Thus, Article 5 of the Constitution provides for the recognition of essential rights and of the international treaties ratified by the nation. Today, there is a complete set of international norms and regulations that have been signed by various nations that, among other things, give legal substance to the study of space as social production, whether through territories, the city and/or housing. Chile has signed several treaties of this nature, but they have not yet been implemented. Perhaps the right to housing is the most important one to ratify among all the purposes we invite you to review in this book.

In 2013, two of the main organisations of settlers in Chile, the MPL and ANDHA Chile, wrote a letter to the National History Prize winner asking him for historical and useful information to move towards a constituent moment. The letter focuses (or elaborates) on the convergence of constituent forces in Chile is inevitable. However, social movements and peoples in rebellion face the same historical threat that has skewed the premature future of our deliberation and postponed the exercise of sovereignty from below. In this constituent path we need to meet with all the forces walking towards a new society, with a new project of a dignified life for all.[15]

As an anticipation of what happened in 2019, the MPL and ANDHA Chile announced that there was a fatigue with representative institutions and the political class, especially the apparent self-representation of the latter, leaving aside the people and their struggles. In this premonitory vision, it was proposed to move towards a constituent moment where the struggle to change the conditions of life would be reflected in a new way of living in the nation. Given that the MPL and ANDHA Chile are movements that fight for decent housing and the right to the city, there can only be empathy for their demands, and one wonders how the urban planners' guild views the possibility of creating a new Constitution.

In Chile after 18 October 2019, the urgency for a new Constitution has become a national priority. However, the architects' guild, especially within the Habitat and Housing Committee of the Architects' Association, have been reflecting for years on the possibility of moving towards a new Constitution,[16] where space is an integral part in order to cover part of its complexity and ensure its correct social allocation. In their report, they propose the importance of taking the opportunity offered by the constituent moment, that is, when society as a whole begins to understand that the Constitution is key to defining the problems of the everyday political-economic structures that govern and which, in the Chilean case, need urgent change. It is a time when society, and not only the elite, is aware of the need for change. From this 2016 missive, a strong emphasis is placed on the importance of human rights represented in the habitat. For this, the social function of land, property, the city and housing are key to unlocking the remaining dictatorial vestiges of the 1980 Constitution. Other aspects are functional in the institutions, to which the Habitat and Housing Committee calls for the state to recover its planning and land-use functions, which have been dismantled and handed over entirely to the market in order to plan on the basis of private property rather than the common good.[17] This incorporates the need to generate new models of governance to manage territories and promote the democratisation of public structures in order to move towards processes of binding citizen participation. Finally, in the search for the common good, environmental rights are fundamental. In an extractivist nation such as Chile, it is essential that the environment be protected to ensure that there is natural heritage, access to water and, obviously, access to a future in the face of climate change. There are also other expert views that call for a profound change that is reflected in the Constitution. In this respect, Miguel Lawner, winner of the national prize for architecture, proposes reviving the provisions of Law 16.615 of 18 January 1967, which states that the common good should be prioritised over the private good, with emphasis on natural resources and productive assets. With regard to private property, the role of the state is specified: "It will also promote the appropriate distribution of property and the constitution of family property".[18]

Such constitutional changes would allow the state, not the market, to organise cities for the common good. The need to give the state new powers to move towards the rule of human rights and the common good is not trivial, but without constitutional change, this will be virtually impossible.

Many of the reflections presented in this book are indirectly a critique of the strong private property regime enshrined in the current Political Constitution. In it, the city, housing and space are invisible. We believe that the following background will make it clear that there is an urgent need to incorporate space as a social product into our fundamental charter.

Notes

1 Atria, Fernando Atria et al., *El otro modelo* (Santiago: Random House Mondadori, 2013).
2 Solimano, Andres Solimano, *Chile and the Neoliberal Trap: The Post-Pinochet Era* (Cambridge: Cambridge University Press, 2012); Dante Contreras y Ricardo Ffrench-Davis, «Policy Regimes, Inequality, Poverty and Growth: The Chilean Experience, 1973–2010», *Unu-Wider*, 2012, www.wider.unu.edu/stc/repec/pdfs/wp2012/WP2012-004.pdf.
3 *OECD Urban Policy Reviews, Chile 2013*, OECD Urban Policy Reviews (London: OECD Publishing, 2013), https://doi.org/10.1787/9789264191808-en.
4 Lefebvre, Henri Lefebvre, *La producción del espacio* (Madrid: Captian Swing, 2013), https://doi.org/10.1017/CBO9781107415324.004.
5 Bourdieu, Pierre. *The Social Structures of the Economy*, 1st ed. (Cambridge: Polity Press, 2005).
6 Political Constitution of the Republic of Chile, Article 24.
7 *Ibid.*
8 Lefebvre, Henri Lefebvre, *La producción del espacio* (Madrid: Captian Swing, 2014).
9 Daher, Antonio Daher, Territorios de la financiarización urbana y de las crisis inmobiliarias, *Revista de Geografia Norte Grande* 1, no. 56 (2013): 1–19.
10 Boano, Camillo Boano y Francisco Vergara-Perucich, *Neoliberalism and Urban Development in Latin America* (London and New York: Routledge, 2017).
11 Vergara, Jorge Eduardo Vergara, Verticalización. la edificación en altura en la región metropolitana de Santiago (1990–2014), *Revista INVI* 32, no. 90 (2017): 9–49.
12 Paulsen, Alex Paulsen, Negocios inmobiliarios, cambio socioespacial y contestación ciudadana en Santiago Poniente. El caso del barrio Yungay: 200–2013, *La ciudad Neoliberal, gentrificación y exclusión en Santiago de Chile, Buenos Aires, Ciudad de México y Madrid* 2014, 75–98.
13 Hidalgo Dattwyler, Rodrigo Hidalgo Dattwyler, Alvarado Peterson Christian Voltaire, y Daniel Santana Rivas, La espacialidad neoliberal de la producción de vivienda social en las áreas metropolitanas de Valparaíso y Santiago (1990–2014): ¿hacia la construcción idelógica de un rostro humano?, *Cadernos Metrópole* 19, no. 39 (2017): 513–35.
14 Burgos, Soledad Burgos et al., Residential typologies in Chilean irregular settlements with precarious housing conditions, *Revista Panamericana de Salud Publica* 29, no 1 (2011): 32–40, https://doi.org/dx.doi.org/S1020-49892011000100005.
15 MPL y ANDHA en Gabriel Salazar, *El poder nuestro de cada día. Pobladores. Historia. Acción popular constituyente* (Santiago: Lom Ediciones, 2016), 12.
16 The Chilean Association of Architects and the Habitat and Housing Committee launch the campaign New Constitution: Human Dignity in Territory and City. http://colegioarquitectos.com/noticias/?p=12184.
17 Comité Hábitat y Vivienda, *Dignidad Humana en el Territorio y la Ciudad* (Santiago: Lom Ediciones, 2016).
18 Lawner, Miguel Lawner, *¿Qué hacer?* (Santiago: Carta personal, 2019).

3 Cost of living

We argue that political economy focuses on the problems of inequality resulting from social relations. In this case, we understand that housing is not just a physical space, but a shelter in which people develop their life projects. These life projects will be more or less successful to the extent where the economy does not compress their plans. This is critical in the case of Chile, given that the price paid to live in its territory is a price that most people do not pay. The cost of living in Chile is an issue that was one of the main slogans of the popular movement arising from the social outburst after 18 October 2019. There is no mystery here—there are enormous difficulties for families to make ends meet with the wages paid in the country. These salaries do not cover the welfare that the neoliberal model proclaimed. Under Milton Friedman's monetarist theory, the circulation of money was supposed to adjust to demand. This theory, however, is fragile in a context of high inequality such as the Chilean reality. In concrete terms, money circulates, but the highly unequal distribution of income means that some households achieve well-being only through purchasing power, while others, even with state subsidies, cannot reach minimum levels of well-being. In the Chilean case, once this flaw in the model was detected, the rate of bankarisation increased, and massive indebtedness with credit cards of low financial quality and high interest rates began to subsidise the purchase of basic goods for many households. In Chile, it is not uncommon to see people buying food with credit cards. Basic consumption began to enter financial frameworks in a sustained process of the financialisation of everyday life. Between 2006 and 2015, the number of credit cards per capita in Chile went from 0.25 to 0.71. In general, the problem of the cost of living has been far from the public policy agenda as a structural problem of a highly unequal neoliberal society. While there have been targeted policies, these generally attack the specificities of the high cost of living: subsidies for certain medical treatments, cash bonuses for households in winter for some lower-income households, bonuses for families with children when they

DOI: 10.4324/9781003348771-3

start school, etc. The reproduction of inequality and indebtedness as a palliative mechanism of such inequality was out of public discussion for years. Only after a social outburst, the pandemic and the process of constitutional change has this problem been brought to the national political economy agenda as a priority. Certainly, the negligence of the ruling class on this issue was one of the reasons the Chilean citizenry lives in a condition of indignation. Undoubtedly, this is happening elsewhere in the world where neoliberalism has also imbued everyday life with its monetarist flavour, painting all social relations with the colour of the profitability of decisions.

In Chile, centralism has been a key factor in the reproduction of inequality and an example of this is that much of the discussion on the cost of living has focused on the city of Santiago—comparing it, for example, with other world capitals. In the regions, however, the problem is not unrelated, although its magnitude is not entirely clear. The scarcity of official data does not help. In this chapter, we contribute to the debate with a question that we believe should be the starting point for political economy studies: How much does it cost households to pay the costs of living in order to achieve a minimum standard of living? In this sense, the data collected allow us to contribute to measurements of inequality differentiated by territory, but also allow us to consider, from the microeconomics of households, how the price of housing is critical for well-being.

The study presented in this chapter presents an analysis of living costs in the main Chilean cities, using data from the 2017 National Socioeconomic Characterisation Survey (CASEN). This survey presents some sampling problems when trying to work at the scale of smaller communes, but it works better when used for large urban populations. For this reason, a study based on regional capitals is carried out. The initial reference is the autonomous income of the household according to the socio-economic decile. From this income, the average monthly expenditure on housing made by households is deducted (including rent or dividend, as the case may be) and then food expenses and the average monthly value of transport are deducted. Only cases where households pay a dividend or rent have been taken, in order to improve the representativeness of the sample data in terms of their monthly housing expenditure.

In concrete terms, this study evaluates the purchasing power of households in relation to the basic costs of daily life taken from each urban area by regional capitals, including the metropolitan area of Santiago. Depending on the city, living costs change, but this is not necessarily registered in the instruments for defining public policy, due to the centralist logic of our political-administrative apparatus. Cost changes are mainly registered in housing and transport. We assume that food costs also vary, although we have decided to use only the data from the National Statistics Institute (INE) for the basic food basket, registering variations according to the composition of the household, that is, according to how many people live there.

The results indicate the expenditures in the various regional realities, given that indicators such as the Consumer Price Index (CPI) or the Family Budget Survey[1] do not end up identifying how much it costs to live in Chile by communes or in other more precise spaces. In a national scenario of outrage against the effects of the monetarist model, where monetary circulation and consumption are assumed to be the economic drivers, knowing the costs of living accurately can help better inform the malaise. This also informs political economy analysis as households' ability to cope on a day-to-day basis and, in the face of difficulties in doing so, to understand some of the general malaise in the country against the political-economic model.

Defining indicators to assess the cost of living is a topic that has attracted interest in the micro and macroeconomic literature. In Chile, there is a tendency to use advanced statistical and econometric mechanisms to identify variations that, in practice, are difficult for people in general to read, given that they necessarily reflect the household economy. This is evident in Central Bank reports or scientific publications, which—although very necessary—are limited and partial in their downstream dissemination to everyday life, harming the dissemination of the results to society as a whole. Thus, something as simple and informative as the relationship between household income and fixed expenditure remains relatively absent for the majority of the population, when this is precisely the economic factor that has the greatest impact on well-being. This is a rather local problem, given that in other realities and in a vast literature, these discussions are better informed and openly expressed, seeking a more universal understanding.

An exhaustive study on price dynamics in relation to the cost of living and inflation was developed in the United States,[2] whose approach to the problem was used for the production of this study. This reference proposes a comparison between different geographical entities (a country, a city, a region) in the same period of time, favouring the use of official panel-type data sources in order to work on the basis of integrated information that allows for future comparisons, generating a useful instrument to initiate a follow-up. In the case of Chile, the surveys are not panel surveys, which is a disadvantage. In addition, it is important to consider transport in the evaluation of parity purchasing power. On the other hand, as Paredes and Aroca[3] argue, there are important variations in living costs within the country, depending on regions and specific cities.

Certainly, developing cost of living studies between regional capitals makes it possible to determine territorial inequality relations, as long as these studies are carried out on the basis of the same type of currency.[4] In this context, the question to be asked is the following: What percentage of income do households in different localities pay for access to housing?[5] In other words, we are talking about a problem of affordability.[6] This concept is understood as the ability of households to afford housing and other basic goods and services based on income as a means of subsistence.[7] In other words, it is not a linear and universal measure, but relative to each particular context of the households analysed, given that incomes within territories are not uniform.[8] In other words, in order to compare affordability of daily life by cities, it should not be measured in monetary outcomes but in income-expenditure ratios. The literature also emphasises the importance of identifying income segments of the population to determine public policies that define what is considered affordable and meeting certain basic housing and urban living standards that do not tend to segregate social classes, so that public spending can better target where to inject resources in search of equity.[9] This could be understood under the logic of a subsidiary state, as Chile's has been for the last 40 years. Housing affordability studies seek to understand how to ensure that the population has a decent floor and roof under which to live, which ideally should also incorporate *nonhousing items*, as suggested by Stone.[10] Based on these recommendations, we have had to make some decisions based on the information available in Chile, excluding clothing, health and childcare. This is because the source of financing for these factors is highly variable in the case of health (very significant difference between the public and private sectors), childcare (public, subsidised and private institutions) and due to the lack of data on clothing expenditure at the community level.

David Hulchanski defines six methods of analysis to study the relationship between income and housing expenditures: (1) description of housing affordability; (2) trend analysis; (3) study of the provision of public housing and subsidies according to socio-economic levels; (4) definition of housing needs themselves for the development of public policies; (5) prediction of households' ability to pay rent or mortgage; and, finally, (6) analysis of the decisions consumers make when renting or buying a property.[11] To these ways of analysing affordability, it would be necessary to add Stone's proposal to move towards a more comprehensive analysis and thus transform the methods of housing analysis to a study of affordability of everyday life, including *nonhousing items*.

Based on these theoretical contributions, we propose the development of an exploratory study to describe the affordability of daily life in the main Chilean cities, taking the costs associated with housing (rent and dividends), monthly public transport costs, basic food basket and autonomous income per household according to the CASEN 2017 survey (transforming values to UF[12] to absorb the differences in price variation between the different years to be reviewed) as the main references. For the use of the CASEN

2017, which is not representative in all communes, we have used data only from regional capitals and some conurbations to increase the degrees of representativeness.

The cost of living in Chile seems to be hidden behind a set of macroeconomic figures that indicate that the country has experienced a bonanza in the last decades that spills over (or "trickles down", in colloquial terms to explain the theory known as *trickle-down economics*[13]) to the whole population. Even recent Gini coefficient results indicate that income distribution has improved. This is a mirage that is diluted when one changes scale and, in a fractal exercise, reviews the state of the household economy in specific sectors of the population. It is in this sense that the data presented here are quite clear and hard and at the same time, understandable for the majority, with the objective of pondering from the daily experience of households whether the big figures make sense with reality.

To illustrate the inequality in household economy, the main cities of Chile are taken, the cases are weighted by the communal expansion factors suggested by the Ministry of Social Development for the work with the CASEN 2017, the sample is segmented by regional income decile and the following data is extracted: corrected autonomous household income, imputed rent and number of people per household. Then, information is collected on the cost of public transport fares for each regional capital and multiplied by two persons and 40 trips, in order to have a reference value of a minimum monthly expenditure. Finally, the average number of people per household according to income decile is taken and multiplied by the price of the last basic food basket issued by INE, in order to know the basic expenditure of households to meet the minimum calories that a person needs to subsist. The monetary values obtained (in Chilean pesos) are transformed into UF of 2017 for the data obtained from the CASEN 2017 survey and into UF of 2019 for the month of January, with the rest of the data to be analysed.

In Chile, the CPI assesses changes in the average price of the cost of the basic basket of goods and services that tends to represent the average monthly expenditures of a typical Chilean family. That is to say, the assumption is that the basket is a measure of how much the cost of living increases in cities and determines the changes in affordability as measured by inflation. But in Chile, the CPI has a fundamental contradiction, in that it is geographically blind. This adds to the blindness on some high-cost staples, which are under-represented in the samples we have had access to. Obviously, the most severely under-represented factor we found, and therefore the most worrying, is housing, which accounts for less than 12% of the basket. Official data express this difference. While the price of housing in Chile increased by 59% between 2008 and 2016, the CPI varied by only 34% in the same period.[14] However, in the same period, the variation in the price of housing in the northern part of the country increased by 82%. Thus, in practice, housing price variations are not faithfully recorded by the main public mechanism that seeks to register the cost of living in the country, and this can affect wage adjustments, inflationary processes or the targeting of public policies. From this point of view, the high level of indebtedness in Chile[15] could be a consequence of the fact that the cost-of-living measurements do not accurately represent the purchasing power of households in the market, which is why they resort to credit to cover basic expenses. On the other hand, public policy is unable to recognise the real variations in the price of basic consumption factors which, in the hands of the market, fail to be distributed equitably and with a sense of economic justice.

For the analysis, data have been considered for the main Chilean cities by region, including conurbations and potential metropolitan areas, namely, Arica, Iquique and Alto Hospicio, Antofagasta, Copiapó, La Serena and Coquimbo, Greater Valparaíso, Rancagua and Machalí, Talca, Greater Concepción, Temuco, Valdivia, Puerto Montt, Coyhaique, Punta Arenas and Greater Santiago.[16]

At the national level, households spend on average at least 79% of their income on housing, transport and food. However, when this data is broken down by socio-economic deciles, it can be seen that while the richest decile allocates only 15% of household income to paying these fixed costs, the first and second decile—with the lowest incomes—allocate 321% and 112% of monthly income, respectively (Figure 3.1). This is the same as saying that they do not have sufficient purchasing power to monetarily cover the basic costs of living in the regional capitals, but still live there.[17] A significant percentage of the population also spends a large part of their income on living costs, which in no way reflects a healthy household economy.

According to Warren and Tyagi,[18] there is an optimal situation rule for knowing whether the household economy is healthy by relating income against fixed expenses. In the United States, it is known as the 50/30/20 rule applied to household income. That is, 50% of monthly income should go to fixed expenses, 30% to leisure and 20% to savings or investments. Following this rule, Table 3.1 marks in grey those deciles in cities that allocate more than 50% of income to fixed expenses. After applying this criterion, it is clear that the 1st to 4th deciles have significant difficulties in meeting monthly living costs from disposable income, while the 5th decile is not far from 50% either. On average, the most expensive cities to live in from the perspective of affordability of these fixed costs for the lowest income groups are La Serena, Temuco, Talca and Valdivia. Although these are only averages, the results show that the variation in living costs between regional

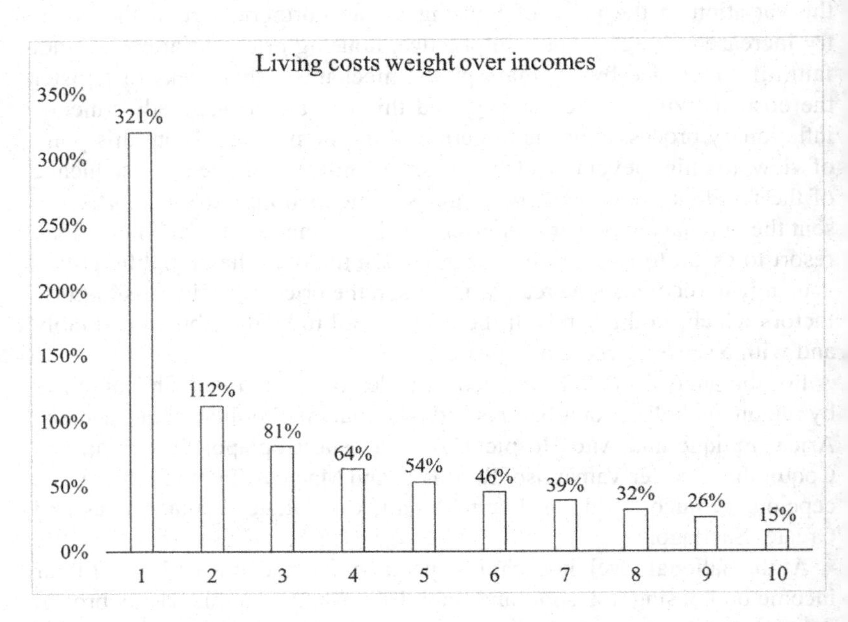

Figure 3.1 Evaluation of basic monthly expenditures as a percentage of household income by decile.

Source: Own elaboration with data from CASEN 2017, INE and researchers.

capitals has important differences that require revisions on the economic health of the particular households in each case. Given the existence of regional offices of the INE, it is not explained why local cost-of-living indicators are not constructed. On the opposite side of this study, the least expensive cities for the first decile are Coyhaique, Santiago, Punta Arenas and Antofagasta. Certainly, inequality between cities is also notorious. However, when looking at the high-income segments, it is observed that the highest decile has more equality in its living costs in relation to income between cities, showing a potential territorial equity that is also segregating as far as this particular study is concerned.

From the data, it is possible to interpret an apparent possibility of mobility between deciles for segments VII, VIII and IX, given that they have fixed expenditures over similar income, which could indicate that migration between these cities for these socio-economic segments could take place more easily, even allowing them to move up the socio-economic groups. This could be cross-checked with social mobility studies applied to migrants in these segments.

Table 3.1 Evaluation of the cost of living by income decile at the national level by regional capitals. Those that spend more than 50% of income on fixed expenses are highlighted in grey.

Region	*I*	*II*	*III*	*IV*	*V*	*VI*	*VII*	*VIII*	*IX*	*X*
Arica	241%	110%	81%	69%	60%	49%	44%	37%	29%	17%
Iquique	165%	91%	70%	58%	50%	44%	38%	31%	28%	17%
Antofagasta	206%	98%	71%	59%	46%	41%	34%	26%	22%	13%
Copiapó	364%	119%	87%	65%	55%	46%	36%	30%	24%	17%
La Serena	339%	143%	97%	75%	69%	60%	52%	40%	35%	19%
Gran Valparaíso	209%	87%	67%	53%	49%	43%	34%	31%	26%	16%
Rancagua	211%	95%	74%	60%	49%	44%	37%	32%	27%	17%
Talca	257%	120%	94%	69%	57%	50%	42%	39%	28%	14%
Gran Concepción	273%	118%	89%	70%	58%	48%	41%	33%	27%	15%
Temuco	1170%	157%	99%	81%	67%	55%	49%	39%	29%	15%
Valdivia	384%	133%	96%	73%	60%	55%	45%	37%	30%	15%
Puerto Montt	406%	131%	84%	70%	60%	47%	35%	30%	24%	14%
Coyhaique	211%	95%	68%	56%	46%	35%	28%	23%	17%	9%
Punta Arenas	210%	90%	67%	55%	46%	39%	31%	30%	23%	12%
Gran Santiago	175%	91%	68%	55%	45%	39%	33%	26%	22%	13%

Source: Own elaboration

The results are illustrative of the country's socio-economic inequality, but they also show the significant territorial variability of these inequalities. In Chile, living costs are measured with a single CPI, without values disaggregated by region or territory. This drastically reduces the accuracy of inflation measurements; at the same time, this shortcoming hides the precariousness experienced by some communities. In addition, it is important to increase the number of socio-economic assessments in order to have more frequent data to inform the economic progress of households. It is also important to improve the representativeness of the data for all surveys at the community level, increasing the samples and applying geographical factors to better understand how the population is distributed in the territories. Although improving the collection of social data implies a high cost for the state, it is also costly not to know how unequal society is. Inequality is sowing social outbursts such as those of 2006, 2011 and 2019. If the state is blind to socio-economic problems, policy responses will be ineffective.

In terms of specific outcomes, lower-income households develop subsistence strategies in the face of the evident financial insolvency they face every month. Phenomena such as shantytowns, indebtedness, slums and informal housing are existing responses to make ends meet in Chile. This is worrying. A matter that was postponed for years by the political economy in Chile was ways to move towards redistribution of wealth and means to increase household incomes. The 1980 Constitution prevents the state from playing an active role in the market, including the labour market. For this reason, wages were mainly adjusted to a CPI that does not really measure the costs of living in Chile and with that, precariousness increased.

One of the problems of the CPI measurement is the undervaluation of the cost of housing, which represents only 8% of the basket of products that make up the CPI. This is despite the fact that housing is the main monthly fixed cost factor affecting the cost of living. In Chile, housing is a basic but deregulated consumer good, whose price has been rising sharply for some years now, squeezing family budgets, reducing household purchasing power and segmenting cities into rich and poor neighbourhoods, establishing a face of inequality that is drawn on Chile's urban maps. The articulation of the political economy with social inequality requires housing to advance the integral complexity of the problem.

Notes

1 This survey presents two geographies for the data study: Greater Santiago and the rest of Chile, as if living in Antofagasta, Punta Arenas or Iquique is economically equivalent to Valparaíso, Temuco or La Serena.
2 Stephen G. Cecchetti, Nelson C. Mark, y Robert Sonora, Price Index Convergence among United States Cities, *International Economic Review* 43, no. 4 (2002): 1081–99.
3 Paredes, Dusan Paredes y Patricio Aroca, Metodología para Estimar un Indice Regional de Costo de Vivienda en Chile *, *Cuadernos de eConomía* 45 (2008): 129–43, https://doi.org/10.4067/S0717-68212008000100005.
4 Kakwani, Nanak Kakwani y Robert J. Hill, Economic Theory of Spatial Cost of Living Indices with Application to Thailand, *Journal of Public Economics* 86, no. 1 (2002): 71–97, https://doi.org/10.1016/S0047-2727(00)00174-2.
5 Amoako, Clifford Amoako y Emmanuel Frimpong Boamah, Build as You Earn and Learn: Informal Urbanism and Incremental Housing Financing in Kumasi, Ghana, *Journal of Housing and the Built Environment* 32, no. 3 (2017): 429–48, https://doi.org/10.1007/s10901-016-9519-0; Gilberto M. Llanto, Shelter Finance Strategies for the Poor: Philippines, *Environment and Urbanization* 19, no. 2 (2007): 409–23, https://doi.org/10.1177/0956247807082821.
6 Stone, Michael E Stone, Housing Policy Debate What is Housing Affordability? The Case for the Residual Income Approach What Is Housing Affordability? The Case for the Residual Income Approach, *Housing Policy Debate* 17 (2010): 37–41, https://doi.org/10.1080/10511482.2006.9521564; Tony Roshan Samara, Anita Sinha, y Marnie Brady, Putting the "Public" Back in Affordable Housing: Place and Politics in the Era of Poverty Deconcentration, *Cities* 35 (2013): 319–26, https://doi.org/10.1016/j.cities.2012.10.015; J. David Hulchanski, The Concept of Housing Affordability: Six Contemporary Uses of the Expenditure to Income Ratio, *Housing Studies* 10, no. 4 (1995), https://doi.org/10.1080/02673039508720833; Emma Mulliner y Vida

Maliene, Criteria for Sustainable Housing Affordability, *Environmental Engineering* (2011): 966–73; Peter D. Linneman y Isaac F. Megbolugbe, Housing Affordability: Myth or Reality?, *Urban Studies* 29 (1992): 369–92, https://doi.org/10.1080/00420989220080491.

7 Stone, M. E. "Housing Policy Debate What is housing affordability? The case for the residual income approach What Is Housing Affordability? The Case for the Residual Income Approach, 17 (September 2012), 37–41." (2010).

8 Redding, Stephen Redding, Spatial Income Inequality, *Swedish Economic Policy Review* 12, no. 1 (2005): 29–55.

9 Stone, *op. cit.*

10 *Ibid.*

11 Hulchanski, J. David. "The concept of housing affordability: Six contemporary uses of the housing expenditure-to-income ratio." Housing studies 10.4 (1995): 471–491.

12 UF is an indexed financial value measured in Chilean pesos which is adjusted to the inflation of the Chilean economy, whose price is defined by the Central Bank and used for mortgage operations. In general, housing and subsidees are valued in UF in Chile.

13 The so-called "spillover" or "trickle-down" effect (as it is known in Chile) belongs to neoclassical economic theory and proposes that taxes or regulations applied to wealth or big business in a country should be reduced as a way of favoring investment in the short term and hoping, as a result, that this would improve economic conditions for the rest of the population in the long term. In Anglo-Saxon literature, this concept is usually associated with the economic policies of the governments of Ronald Reagan and Margaret Thatcher in the 1980s in the United States and the United Kingdom, respectively, although in Chile, it found particular resonance in justifying the change of the economic model proposed by the civil-military dictatorship during the same period.

14 Data as reported by the Central Bank in its *online* comparable statistics site is available at this link: https://si3.bcentral.cl/Siete/secure/cuadros/home.aspx

15 Chovar, Alejandra, and H. Salgado. "¿ cuánto influyen las tarjetas de crédito y los créditos hipotecarios en el sobre endeudamiento de los hogares en chile." Banco Central de Chile, Vol. 8 (2010). Available at: https://www.bcentral.cl/documents/33528/133585/bcch_archivo_140212_es.pdf/16a59094-1e07-9f2e-aa2b-63c4aca06f4b?t=1573288203160 (accessed 22-11-2022)

16 Number of cases used for analysis by regional capitals: Arica: 7,663; Iquique + Alto Hospicio: 9,462; Antofagasta: 5,473; Copiapó: 3,757; La Serena + Coquimbo: 5,217; Valparaíso + Viña del Mar: 5.199; Rancagua + Machalí: 3,902; Talca: 2,086; Greater Concepción: 6,482; Temuco + Padre Las Casas: 4,744; Valdivia: 4,765; Puerto Montt + Puerto Varas: 3,310; Coyahique: 3,151; Punta Arenas: 5,809; Greater Santiago: 31,868.

17 In an interview on CNN Chile (29 October), social leader Soledad Mella asked guests if they knew what it was like to live in Chile on the minimum wage ($301,000), to which she said she could explain in detail. Unfortunately, there was no time to explain how, and this indicator shows that it is a real mystery hidden behind the brutal inequality that the country is experiencing.

18 Warren, E., and Warren-Tyagi, A. *All Your Worth* (New York: Free Press, 2005).

4 House prices in Chile

For decades, Chile has fostered the aspiration to be a country where everyone has access to affordable housing. This has implied important efforts on the part of the public sector, with special emphasis on lower income sectors. However, despite many years of progress in reducing deficits, as of 2022, we can see that the structural housing deficit has not been significantly reduced since 2000. The Zero Deficit initiative has established that the deficit is close to 600,000 units, much higher than in 1998. On the other hand, TECHO, through its Centre for Social Research, has stated that the number of slums is beginning to resemble that of 1985, after the earthquake that year. Housing is a necessary commodity for survival and life. If housing is in crisis, so is the society. In this case, the crisis stems directly from the individualistic principles of neoliberal political economy.

In mid-2019, there was some consensus among private, public and academic actors that high housing prices were leading us towards a serious social situation in the medium term. Similar concerns were expressed by the Central Bank of Chile about the increase in the number of people using housing as an investment vehicle. The crisis was looming and, in our view, was one of the causes of the social outburst of October 2019. This assertion is supported by research results from the Centro Producción del Espacio of the Universidad de Las Américas, publications of the Instituto de la Vivienda of the Universidad de Chile, the Instituto de Economía Aplicada Regional of the Universidad Católica del Norte and the Instituto de Estudios Urbanos y Territoriales of the Pontificia Universidad Católica de Chile. The results indicate that there are various causes for the price increase, which are complex and multifactorial, mainly related to an imperfect, deregulated and non-transparent market. Although some diagnoses converge, difficulties arise in the search for solutions. This is mainly because there are ideological agendas and political and economic class agendas behind the discussions on this problem. The clash of these agendas often results in contradictory proposals. While actors linked to the

DOI: 10.4324/9781003348771-4

real estate business argue that the high price is due to a shortage of land and excessive regulations, others, such as some directors of the Architects' Association, support the idea that strict urban planning regulations mean that investors are uncertain about projects, making them more expensive. Academics, meanwhile, argue that one of the main perversions of the value of housing use is that it has been used as an investment vehicle by the wealthiest sectors, with the aim of generating future economic sustenance from rents. The latter, which may be very good for these sectors, is putting a huge financial strain on a large part of the population. We will look at this particular phenomenon in more detail below.

This chapter contextualises the most reiterative diagnoses and critically reviews the most frequent explanations for the high price crisis in order to understand the debate and provide the reader with an overview of how we interpret the causes of the problem. In addition, this chapter will allow us to reject some hypotheses on the basis of the empirical evidence we are aware of. The clearest representation of this crisis is the deficit of 600,000 housing units[1] for a period in which nearly 3,600,000 units were built. There are not problems of housing production, but of distribution, which poses a task for the exercise of political economy.

a Hypothesis 1: due to soil scarcity

Arnold Harberger is one of the ideologists of the Chilean neoliberal urban model. His academic vision was fundamental for the creation of the National Urban Development Policy of 1979, which promoted the market as the organiser of the city. Among his theories applied in Chile, he emphasised the relationship between the urban boundary, land scarcity and land prices. In his vision, the urban limit is a fictitious line that turns land into a scarce good and therefore increases its price, which ends up having an impact on the price of housing, making it difficult for the majority of the population to access it.[2] Since then, the issue of land scarcity has been a recurrent argument among those who value the contribution of Harberger and the *Chicago Boys* to urban development. However, empirical evidence indicates that the phenomenon was exactly the opposite, and as urban land was liberalised, the process of concentration of property ownership in the hands of a specific group of controllers who could buy large tracts of land and accumulate them until they decided whether to exploit them for real estate or sell them at the best possible price, in a speculative phenomenon, began. Harberger's theories,[3] as well as those of the National Urban Development Policy, were challenged a few years after their implementation,[4] but the convenience of his arguments is still used by actors in the sector to explain the housing price problem.

This theory is best wielded by the real estate developers themselves, who from the visible platform of the Chilean Chamber of Construction (CChC) influence the opinion of representatives and authorities with little urban planning training and little knowledge of urban literature in Chile, who naively adhere to these principles uncritically, without even considering the fact that the CChC—as a trade association—could use arguments that, precisely, will improve business opportunities and not necessarily promote a holistic view. Therefore, the first hypothesis discussed here is that the high price of housing is largely due to the scarcity of urban land, where, in addition, the little land available is very expensive for business.

But is there such a shortage? Data from the National Institute of Statistics show that in 2018 alone, the area authorised for new housing construction increased by 7.4%, in a trend line of steady increase since 1992. On the other hand, in 2018, real estate sales increased by 10.3% compared to 2017, and the sales of new housing in the Metropolitan Region reached 31,942 units, which added to the data of the steady increase in permits, which would indicate that the real estate industry is not at a standstill, but in sustained expansion for more than two decades. Given this volume of production, then, how does the land shortage argument hold up?

One of the people who has collected loads of data on the land market is the economist Pablo Trivelli. He is an internationally recognised expert with a clear academic profile who, in a conference given at CEDEUS in 2017, indicated that there would not be a lack of land to build new housing[5] given that, according to his data, in the most central sector of the Metropolitan Region, there are 555 hectares available for construction. According to our estimates, this land availability would be enough to build about 500,000 housing units if this land were used to build eight-storey buildings, which covers much more than the region's deficit.

On the other hand, toctoc.com, a company dedicated to monitoring the land market, indicated in a 2019 study that in Greater Santiago, there are 3,405 hectares available in communes with good urban attributes, where housing could be developed.[6] Again, the idea of a shortage seems questionable. Meanwhile, in 2018, data from the Real Estate Registry of Santiago and San Miguel indicate that during that year, 4,961,961 UF were transacted on sites with building potential, which would accommodate the construction of more than 500 eight-storey buildings.

Another organisation that has been monitoring the problem of accessibility to residential housing for decades is Fundación Vivienda. According to its data, in the Metropolitan Region, there are 3,152 hectares with good location and densification potential that could be used to improve accessibility to housing for more than 700,000 families, much more than what Santiago needs.[7] In addition, the researcher Ivo Gasic argues that the scarcity of land is artificial and has more to do with the pressure imposed by actors with a lot of capital who buy land to have a future investment reserve, without exploiting it immediately.[8] In other words, in addition to the fact that there is no such shortage, a significant amount of land belongs to individuals or institutions that hope to exploit it when the prices are in line with their income expectations or specific economic needs.

A review of the annual reports of large real estate companies confirms the hypothesis of land accumulation because they already have land banked for the development of future projects. For example, in its 2018 annual report, Ingevec indicates that it has a bank of 26 plots of land, the same amount that Paz Corp. declares. For its part, the company Socovesa bought 12 plots of land in 2018 and Moller Perez-Cotapos bought 17, at an average value of 13 UF/m^2. Socovesa's annual report states that the company has "land for real estate development equivalent to 4 years of sales".[9] This means that the exploitation of these plots of land is based on the company's plans, on the basis of land capital already created, with no urgency to acquire new land. In this sense, it is also possible to say that the company does not face economic constraints due to a lack of annual productivity, given that it would otherwise develop its extensive land bank more aggressively. In other words, expectations are managed according to the fulfilment of financial targets. An example of this way of using land is Ingevec in the north of Chile, where the demand for housing is high, which is reflected in the drastic increase of slums.[10] However, in their annual report, they bluntly state that "the company currently owns some land in the north of the country, for which development is being postponed, betting on a future economic recovery of the region".[11]

Despite the social urgency in the area, the real estate company holds on to its land in the hope of a better return on investment. The problem, then, does not seem to be the scarcity of land, but rather that its owners are waiting for the right moment to exploit it, ensuring high profits.

Now, based on the data provided by toctoc.com and considering the existence of 3,405 hectares in Greater Santiago, how many dwellings can fit there? We have carried out a large-scale study on the data provided by the company, based on a robust estimation from the urban planning instruments of each commune evaluated. In this exploration, we have analysed each commune according to data from each local ordinance of the corresponding Communal Regulatory Plan, for the case of isolated buildings. We have not considered spacings or attachments, something that should be developed for each particular plot of land to increase the rigour of the results. For this exploration, we have used a 60m^2 flat as a reference, and we have used the norms inscribed in each municipal ordinance that privilege medium or high densification (Table 4.1).

Table 4.1 Projection of housing potential to be developed by commune according to available land.

District	*Hectares*	*Constructibility* (Ratio)	*Minimum plot size* (sqm)	*Potential plots* (N)	*Land occupation allowance* (N)	*Surface allowed to build* (sqm)	*Floors* (N)	*Potential housing* (N)	*Potential inhabitants* (N)
Quinta Normal	467	1.2	1,200	3,892	1	1,440	2	74,720	224,160
Independencia	231	4	800	2,888	1	3,200	5	123,200	369,600
Recoleta	366	2.4	1,200	3,050	0	2,880	6	117,120	351,360
Providencia	166	2.9	1,600	1,038	0	4,640	7	64,187	192,560
Santiago	416	4.8	500	8,320	1	2,400	7	266,240	798,720
San Miguel	458	2	2,000	2,290	1	4,000	3	122,133	366,400
San Joaquín	92	3	1,800	511	1	5,400	4	36,800	110,400
Ñuñoa	253	2.5	2,000	1,265	1	5,000	5	84,333	253,000
Macul	120	1.8	2,000	600	0	3,600	5	28,800	86,400
La Florida	479	2	3,000	1,597	0	6,000	5	127,733	383,200
Vitacura	134	1.6	3,000	447	0	4,800	4	28,587	85,760
Las Condes	116	2	3,000	387	0	6,000	7	30,933	92,800
Lo Barnechea	106	2	4,000	265	0	8,000	6	28,267	84,800
TOTAL	3,404							1,133,053	3,399,160

Source: Own elaboration.

The result is overwhelming for the land scarcity argument, given that in a scenario where current regulatory situations are maintained, the potential for 1,133,053 new houses is not consistent with housing prices. It is possible, then, that real estate companies do not refer to total availability in Greater Santiago, but rather to certain types of land where the sale of housing is more secure, and therefore, less risky. Between 2010 and 2017, in Greater Santiago, authorisations for large-scale new housing were preferentially located around the central area comprising the Alameda-Providencia-Apoquindo corridor, together with Metro stations. Developments in La Dehesa, Maipú and San Bernardo also stand out, but on a smaller scale. Given that the large real estate companies each have their own land banks, when they talk about scarcity, they are talking about buying in certain places with consolidated urban attributes and where high-cost sales are certain.

However, when reviewing transactions in the Santiago Real Estate Registry, in 2018 alone, several real estate companies registered purchases of well-located urban land (close to consolidated Metro lines) amounting to 17,779 m^2, paying an average of 61 UF/m^2 (Figure 4.1). In this purchase of land, it is evident that scarcity has not been the problem. Specifically, what could be verified is that in 2018, the increase in the value of land was significant in relation to previous years, which could make it difficult to

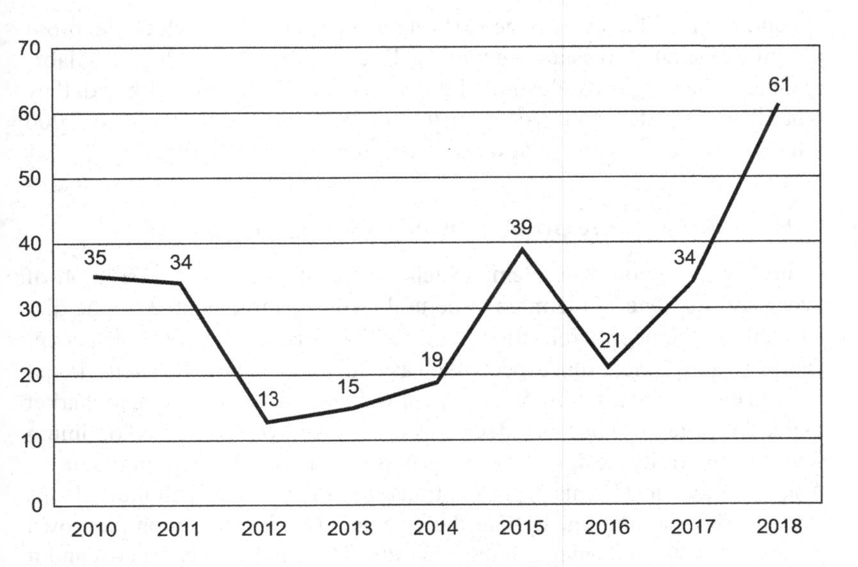

Figure 4.1 Value of land purchased by real estate companies measured in UF/m^2.

Source: Own elaboration based on data from inciti.com.

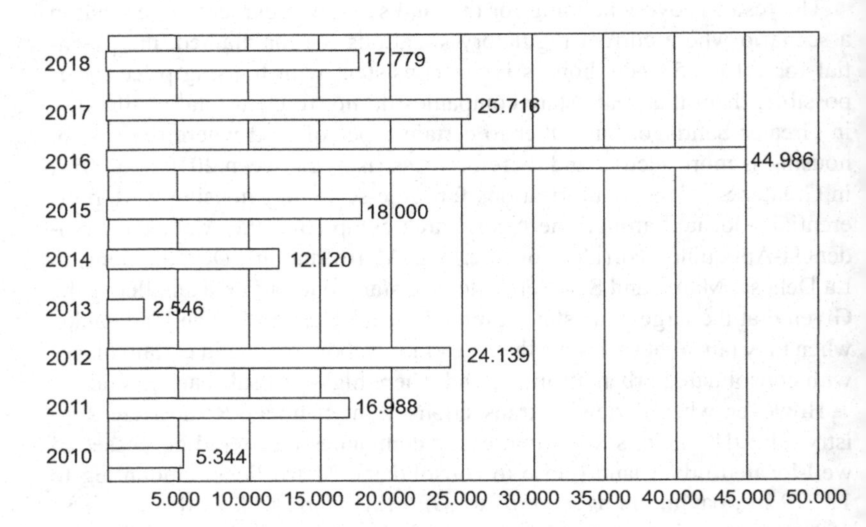

Figure 4.2 Surface area of land purchased by real estate developers in sectors close to the Metro (m^2).

Source: Own elaboration based on data from inciti.com.

expand the land banks of large real estate companies and reduce the profit margins of small real estate companies. Even so, there is still land available in Greater Santiago. To illustrate the size of those 17,779 m^2 marketed, they would be equivalent to the construction of 1,915 new flats for about 7,000 inhabitants,[12] only with the new land acquired in 2018 (Figure 4.2).

b Hypothesis 2: the problem is not housing, it is wages

Chile has a structural problem, which is related to the monetarisation of everyday life. One of the main structural reforms implemented by the dictatorship in the neoliberalisation plan—which was deepened in subsequent years—was to ensure that aspects such as education, health, pensions, housing and even infrastructure development were managed according to market logic. The state itself began to decide its investments on the basis of optimum social profitability designed by the group of economists advising General Augusto Pinochet. In this profound transformation, along with individualisation and social alignment, households began to rely heavily on their own pockets to make ends meet. In other words, living in Chile has a cost and it is not low. This may not be news to a Chilean, but it is tremendously striking for many foreigners who arrive in the country and find that, despite paying for high-cost monthly health insurance (Instituciones de Salud Previsional, also known as Isapres), they have to pay again when they are treated, or see that pensioners must continue working because their pensions are not even enough to pay the rent, or that basic goods such as bread, rice and milk pay VAT; not to mention the absence of public housing programmes to ensure access to a safe roof over their heads regardless of the income of households and individuals. We live in a commodified society like no other on this planet. So, in this context, wages are the main source of survival and one of the few factors that can ensure an adequate quality of life. As Peter Winn says, there are two countries in one, that of the rich and that of the victims of the "Chilean miracle",[13] as the liberal philosopher Friedrich Hayek boldly called Chilean neoliberalism in the 1980s. There is no such miracle for the majority, but the natural progress of a global world that has moved towards situations of significant poverty reductions (Table 4.2).

To break with a chauvinism that is difficult to defend in international frameworks, significant progress in poverty reduction can be observed in political projects as different as Belarus (post-Soviet socialism), Kazakhstan (post-Soviet socialism), Uruguay (Frente Amplio de Izquierda), Costa Rica (social democracy), Thailand (parliamentary monarchy), Kosovo (nationalism) and Argentina (Frente para la Victoria, centre-left). Attributing success in the fight against poverty to an extreme neoliberal model such as Chile's is an arrogant statement that does not correspond to the facts. Even in the context of Latin America, income distribution has not been Chile's strong suit, given that in relation to the reduction of the income gap between 2002 and 2017, it ranks a lowly ninth, as shown in Table 4.3.

Table 4.2 Reduction of the poverty gap to USD 5.5 (2011, PPP) between 2004 and 2015.

1	Moldova	−24.3	14	**Chile**	**−6.1**
2	Ecuador	−17.2	15	Uruguay	−6.1
3	Kazakhstan	−15.5	16	Ukraine	−6
4	Peru	−14	17	Lithuania	−5.4
5	Kosovo	−12.3	18	Costa Rica	−4.7
6	Brazil	−10.7	19	Russia	−4.3
7	Panama	−10.4	20	Latvia	−4
8	El Salvador	−9.7	21	Serbian	−3
9	Thailand	−8.7	22	Malaysian	−2.7
10	Paraguay	−8.1	23	Poland	−1.7
11	Argentina	−8.1	24	United Kingdom	−0.3
12	Turkey	−6.5	25	Slovakia	−0.1
13	Belarus	−6.2	26	Israel	−0.05

Source: World Bank.

Table 4.3 Ranking of reduction in income distribution (GINI) in Latin America 2002–2017.

Rank	*Country*	*Reduction*	*Gini 2017*
1	Bolivia	25.90%	45.3
2	El Salvador	22.40%	39.9
3	Ecuador	18.20%	44
4	Uruguay	17.70%	39
5	Peru	17.60%	44.8
6	Argentina	16.90%	38.8
7	Paraguay	13.90%	50.3
8	Panama	11.70%	50.8
9	Chile	10.50%	45.4
10	Colombia	9.90%	51.1
11	Venezuela	9.80%	37.7
12	Honduras	9.80%	48
13	Brazil	5.40%	53.9
14	Mexico	0.40%	50.4

Source: Prepared by the authors based on ECLAC Poverty Report 2018.

Table 4.4 Adjusted autonomous household income at the national level 2017, with data in UF.

Decile	*Count*	*Mean*	*Median*	*Mean (UF)*	*Median (UF)*
I	1746849	154140	140417	6	5
II	2037150	360743	334167	14	13
III	2160375	492922	480000	19	18
IV	1932893	608169	586000	23	22
V	1856091	726835	700000	28	27
VI	1879288	893008	858333	34	33
VII	1703541	1053195	1019417	40	39
VIII	1604406	1270733	1211666	48	46
IX	1523401	1782836	1737500	68	66
X	1343350	3990942	3203833	151	122

Source: Prepared by the authors based on CASEN 2017.

Looking inwards, then, it seems that our macroeconomic performance and the famous growth are not being well distributed among people. This could be impacting on the ability to access quality housing, considering that we live in a society where monetarism—in terms of the ability of households to circulate money—ends up determining our chances of accessing a good quality of life. This distribution of income is profoundly unequal, and while the Gini coefficient allows us to measure it in the broadest sense, the detail can be seen in Table 4.4.

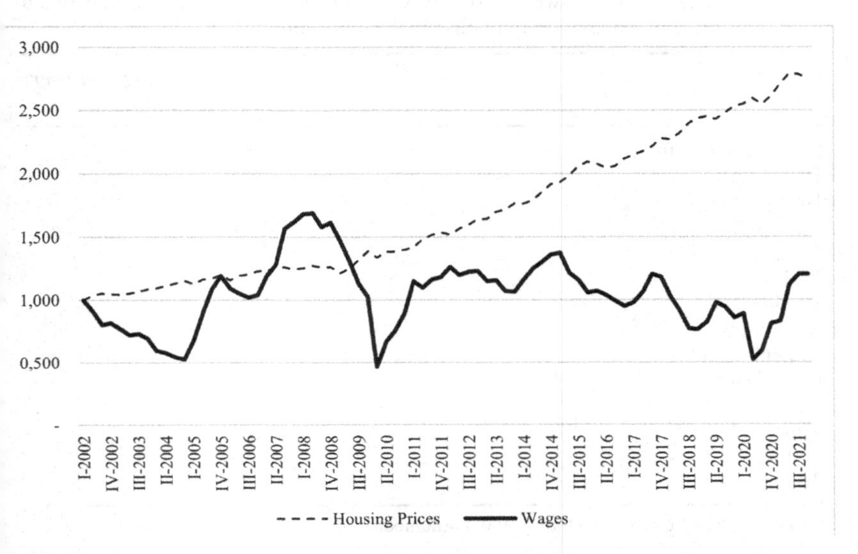

Figure 4.3 Graph showing the relationship between house prices index (blue) and wages index (red).

Sources: www.ine.cl/estadisticas/laborales/ir-icmo and www.cchc.cl/centro-de-informacion/indicadores/indice-real-de-precios-de-vivienda.

In this problem of housing affordability, an interesting aspect emerges when looking at the dynamism of wages versus the price of housing. Although the price of housing is skyrocketing, the stagnation of wages could contribute to the fact that these prices are becoming difficult to pay for more and more people, as it can be inferred from what is shown in Figure 6. The answer here is not entirely clear.

It is clear that we live in cities with developed-country costs, but with third-world wages. An OECD study[14] established that Chile is among the countries with the worst salaries, standing out among the nations with the most precarious jobs in the organisation. As indicated by the Fundación Sol in its multiple studies, wages in Chile do not recognise the value produced in the different jobs, value that in practice is captured by the dominant social sectors.[15] These sectors concentrate a large part of home ownership and in part live from it. The CASEN survey shows that 63% of households in the richest quintile earn a significant proportion of their income from renting urban property.

This unfavourable relationship between wages and housing prices creates serious problems. For Emmanuelle Barozet,[16] it limits the mobility

Table 4.5 Relationship between monthly wage income after taxes and the rental price of a one-bedroom house in the city centre.

City	*Cost of housing rent over average income*
Santiago de Chile	61%
Buenos Aires	56%
Brasília	44%
Washington	42%
London	74%
Paris	52%
Berlin	40%
Melbourne	42%
Copenhagen	45%
Cape Town	58%
Dubai	56%
Vancouver	56%
Seoul	29%
Tokyo	37%

Source: Authors' elaboration based on www.numbeo.com.

of families, confining them to living close to households with similar income levels. Economist Marina Panez[17] points out that the rule that no more than 25% of wages should be spent on housing is rarely observed in Chile, which encourages residential segregation. The comparative table with other cities shows that in Santiago, more than 60% of the salary is spent on rent, which contrasts with Latin American capitals or developed countries (Table 4.5).

But in the face of high housing prices, the market would react by generating lower-value supply to avoid leaving large quantities of flats in *stock*. However, this does not happen; the answer could lie in the purchasing power of households, where a select group of the population with significant purchasing power acquires properties in sectors of the city where they would not go to live, but from where they can extract value by renting these properties to other tenants. This would result in those flats being priced on the basis of the purchasing power of people who do not live in the neighbourhood, in a phenomenon that Kath Hulse and Margaret Reynolds have called *investification*.[18] If this condition is fulfilled in the housing market in Greater Santiago, then the housing market in the city would be adapted to the salaries of the wealthiest and not to those of the neighbours in the neighbourhoods where the property is produced.

c Hypothesis 3: due to regulations

Another recurrent explanation attributes housing prices to problems related to the public apparatus, from unclear regulations to market-distorting policies. For example, for José Miguel Simián,[19] a real estate market expert at the Universidad de Los Andes, the rise in housing prices could be due to restrictions imposed by local regulations and by-laws. His diagnosis is similar to that of Luis Larraín,[20] who points out that height restrictions are "a fashion" that began to operate in well-located districts such as San Joaquín, La Cisterna and San Miguel, and could be the cause of the 30% increase in housing prices.

Those who point against regulations often also complain about the lack of "legal certainty" that would result from Comptroller's Office rulings or court rulings challenging already approved projects. The argument is that further restrictions on the development of projects (e.g., construction stoppages) raise the price of property. The real estate industry has chosen lack of certainty as its second main argument to justify high housing prices, after land scarcity.

The debate around the respect of ordinances is an open wound among architects. The perception that some building permits are approved through questionable interpretations or by taking advantage of loopholes in the regulations leads many to believe that the demand for legal certainty hides, as architect Sebastián Gray[21] says, the ambition to develop real estate businesses without any surprises or scrutiny, even when these may be detrimental to the common good.

In a debate between Humberto Eliash and Miguel Lawner,[22] the latter denounced an open ethical contradiction between the position of the Colegio de Arquitectos regarding the regulatory transgressions of recent housing projects in Estación Central (the so-called "vertical ghettos"). While the Association stood in defence of the developers despite the fact that the Comptroller's Office had established the illegality of the approved works, Lawner demanded a review of the principles and commitments of the Association's architects, in defence of the common good and beauty, both aspects for which the vertical ghettos represent an aggression.[23] Reviewing the threads of this debate on social media, some architect-developers argue that even these disputes over professional ethics also influence legal uncertainty and therefore, housing prices.

But to come to the point, do the regulations we have today generate significant costs that are reflected in higher prices? There are no direct studies that support the hypothesis that the industry maintains. What we have been able to review is that the claim of legal certainty is artful and seeks to validate permits granted by local authorities, even when they conflict with national laws, basic procedures and current regulations. If a construction director approves a building and then the Comptroller's Office challenges the permit for not complying with regulations, perhaps the problem is not the Comptroller's Office but those who designed and approved the project, and the legal uncertainty could arise precisely from designing within the prevailing regulatory frameworks to avoid future challenges. Now, returning to the problem of high property prices, one could argue that greater fluidity in housing production does not guarantee better prices, as the industry has enjoyed—for most of this decade—the legal certainty it now claims, which has not prevented property values from tripling in the same period.

From an ethical and political point of view, moreover, the argument seems spurious, as it seems a way of asking for legal exemptions justified by the structural housing deficit.

d Hypothesis 4: the real estate market is imperfect

Economist Miguel Vargas[24] has found evidence of another possible explanation for the price problem: the existence of tacit collusion in the middle-class housing market. This could be defined as the fixing of a price that is convenient for companies and too high for consumers, which is achieved without reaching an agreement, by adapting to the prices that already exist. In other words, new competitors do not enter with prices that are more convenient for customers. While this collusion is not illegal in Chile, it is an unethical problem and requires better regulation.

Our view takes into account the imperfections of markets. We argue that the high prices we are facing and the lack of supply are caused neither by a lack of land nor by stricter regulations; they are the result of speculation on the housing market, which becomes especially acute in a country where wages have stagnated.

Five key elements underpin property speculation:

1 Encouragement of indebtedness by the authorities. This, through the provision of demand-side subsidies, facilitates the production of homeownership[25];
2 The renunciation of any kind of regulation of housing prices, leaving the free market to act as the sole criterion;
3 The strong private property regime enshrined in the Constitution, which hinders the application of regulations to private-to-private property transactions[26];

4 The significant interest of the financial market in the real estate business[27];
5 And finally, the profitability of real estate investments themselves, which are much more efficient than other financial instruments[28] in capturing long-term fixed capital income. Thus, citizens who do not have the financial capacity to become real estate investors are at the mercy of the imperfect housing market.

Who benefits from this situation? First, large-scale investors, who enjoy insane returns. According to studies that we have been able to develop, in communes of Santiago, returns on investment can be estimated at more than 90%.[29] This means that the final price of housing could be much lower, without losing profitability for investors.

Micro-investors, who buy flats to rent them out, then benefit from them. In a study of 1,496 flats in central Santiago, we identified that the capital gain from buying to rent a flat over a 25-year period could amount to 94% of their current value, while the annual return on these investments could be as high as 11%, which is much better than buying shares on the stock exchange.[30] Knowing the situation of these micro-investors, large investors have already sniffed out the business of sharing profits in exchange for present liquidity and a line of business has been generated based precisely on the future rental expectations of a property.[31]

In the light of these figures, we believe that when housing is no longer seen as a home, leaving behind the concept of the dream of home ownership and beginning to see it exclusively as a commodity, its condition as a good without substitutes and its increase in value over time makes it competitive with other speculative phenomena such as stock market investments, mutual funds or even the PFAs[32] themselves.

This implies that as long as there is a group of citizen-investors willing to pay high house prices, they will not go down. Also, as other companies begin to invest based on the expectation of future capital gains, the problem of housing prices as fixed capital with long-term income will continue to affect the vast majority of the population.

In short, as long as there are no appropriate regulations, it will be the richest who buy housing to rent it out to the less wealthy. This way, housing will continue to be de-profiled and instead of being traded to produce a home, it will continue to be sold as if it were a share package.

We are then faced with a multi-scale phenomenon without a single cause and, for this reason, it is highly complex to work with. In this case, it is essential to debate the different diagnoses, not only those coming from the real estate business. And it is worrying that the authorities and the media validate the hypotheses that emerge from the developers without delving into other aspects.

What threats are on the horizon? We think that a potential stagnation of the housing market is possible when the borrowing capacity of the higher income group reaches its peak. While this is not yet seen as an immediate problem, its potential consequences have already been warned of by the Central Bank in its 2018 financial stability report: the higher income strata recorded a significant increase in mortgage debt[33] related to the acquisition of properties other than the household's home between 2014 and 2017, accompanied by a decline in the debt repayment capacity of most households.

The increase in housing prices is registered together with a 10% increase in Chileans' arrears: today there are 4.6 million debtors in arrears with an average of 1.7 million pesos per person. Let us not forget that it was precisely the relationship between indebtedness and access to housing—"everyone acted as if property prices could rise forever", in the words of David Harvey—that triggered the last great global socio-economic crisis.

Notes

1 Felipe Arteaga of Fundación Vivienda. www.eldinamo.cl/nacional/2018/09/14/deficit-de-viviendas-llega-a-mas-de-497-mil-inmuebles-segun-casen/

2 Encinas, Felipe Encinas et al., Speculation, Land Rent, and the Neoliberal City. Or Why Free Market is Not Enough, *Revista ARQ* 1, no. 102 (2019): 2–15.

3 Harberger, Arnold Harberger, Notas sobre los problemas de vivienda y planificación de la ciudad, *Revista AUCA* 37, no. 1 (1979): 39–41.

4 Donoso, Francisco Donoso y Francisco Sabatini, Santiago: empresa inmobiliaria compra terrenos, *EURE: Revista Latinoamericana de Estudios urbanos y Territoriales* 7, no. 20 (1980): 25–51; Pablo Trivelli, Reflexiones en torno a la política Nacional de Desarrollo Urbano, *EURE: Revista Latinoamericana de Estudios urbanos y Territoriales* 8, no. 22 (1981): 43–64; Antonio Daher, Neoliberalismo urbano en Chile, *Estudios Públicos*, 1990, 281–99.

5 Trivelli, P. (2017). Characterisation of areas with potential for densification in pericentral districts of Santiago. Conference at CEDEUS.

6 El Mercurio, En el Gran Santiago quedan 3.405 ha para construir 1,1 millones de nuevas viviendas, *Diario El Mercurio*, 6 de agosto de 2019.

7 Fundación Vivienda, *Un lugar en la ciudad—6 propuestas de acceso a la vivienda y construcción de ciudades equitativa* (Santiago: Fundación Vivienda, 2018).

8 Gasic, Ivo Gasic, Inversiones e intermediaciones financieras en el mercado del suelo urbano. Principales hallazgos a partir del estudio de transacciones de terrenos en Santiago de Chile, 2010–2015, *Eure* 44, no 133 (2018): 29–50, https://doi.org/10.4067/s0250-71612018000300029.

9 Socovesa, Memoria Anual 2018, *Memoria* (Santiago: Socovesa, 2018), 57.

10 Roof. Slums Monitor. Slums Monitor in Chile, 2018. http://chile.techo.org/cis/monitor/monitor.php#.

11 Ingevec, *Memoria Anual 2018*, SVS (Santiago: Ingevec, 2018), 68.
12 Based on a constructibility coefficient of 4, minimum property division of 1,000 and maximum height of ten floors, with reference to the Plan Regulador Comunal de Independencia in sector ZC1.
13 Winn, Peter Winn, "No Miracle for Us": The Textile Industry in the Pinochet Era, 1973–1998, en *Victims of the Chilean Miracle: Workers and Neoliberalism in the Pinochet Era, 1973–2002*, 2004, https://doi.org/10.1128/AAC.00308-15.
14 OECD, *OECD Employment Outlook 2018*, 2018, https://doi.org/10.1787/empl_outlook-2018-en.
15 Durán, G. and Kremerman, M. Low wages in Chile, 2019. www.fundacionsol.cl/wp-content/uploads/2019/04/Salarios-al-Li%CC%81mite-2017-NV2-1.pdf
16 Sociologist Emmanuelle Bazoret comments on this detail in the following interview: www.latercera.com/nacional/noticia/estudio-revela-creciente-brecha-precio-viviendas-e-ingresos-familiares/642761/
17 Marina Panez develops this idea in the following note: www.publimetro.cl/cl/noticias/2018/12/17/sondeo-publimetro-cuanto-costaria-subsistir-un-mes-en-chile-si-la-vida-se-tratara-de-circunstancias-promedio.html
18 Hulse, Kath Hulse y Margaret Reynolds, Investification: Financialisation of housing markets and persistence of suburban socio-economic disadvantage, *Urban Studies* 55, n. 8 (2018): 1655–71, https://doi.org/10.1177/0042098017734995.
19 Simian at: www.fdi.cl/2018/07/13/estudio-comparo-cuanto-subio-el-precio-de-las-casas-y-cuanto-los-sueldos/
20 Larraín at: https://ellibero.cl/opinion/luis-larrain-viviendas-mas-caras-por-culpa-de-la-regulacion/
21 Gray at: www.economiaynegocios.cl/noticias/noticias.asp?id=561362
22 During this debate, Humberto Eliash was the president of the Architects College of Chile and Miguel Lawner won that year the National Architecture Prize.
23 A summary of this dispute between Lawner and Eliash can be found here: www.biobiochile.cl/noticias/artes-y-cultura/actualidad-cultural/2018/12/21/guetos-verticales-las-disputas-entre-los-arquitectos-miguel-lawner-y-humberto-eliash.shtml
24 Vargas at: http://impresa.lasegunda.com/2016/05/30/A/BM2UIVPV/V82UK80R
25 Hidalgo Dattwyler, R. A.; Paulsen Bilbao, A. G. and Santana Rivas, L. D. Subsidiary Neoliberalism and the Search for Justice and Equality in Access to Social Housing: The Case of Santiago de Chile (1970–2015). *Andamios* 13, no. 32 (2016): 57–81.
26 A full analysis of the right to property in the Chilean Constitution can be found in: Guiloff Titiun, M. Regulatory Expropriation: An Impertinent Doctrine to Control the Imposition of Limits to the Right to Private Property in the Chilean Constitution. *Ius et Praxis* 2, no. 2 (2018): 621–648.
27 Cattaneo Pineda, R. A. Real Estate Investment Funds and Private Housing Production in Santiago de Chile: A New Step Towards the Financialisation of the City? *EURE* (Santiago) 37, no. 112 (2011): 5–22.
28 A demonstration was elaborated by us in a previous *Ciper* article and recently by the international consultancy Collieres.
29 Fresh data in: Vergara-Perucich, F. *Urban Design Under Neoliberalism: Theorising* (Santiago: Routledge, 2019); and in Vergara-Perucich, F. and Aguirre-Núñez, C. *Inversionistification in Latin America: Problematising the Market* (2019). of renting for the Chilean case. *Habitat y Sociedad* 1(12).

30 These data will be published in December this year in: Vergara-Perucich, J. F. and Aguirre, C. Inversionistification in Latin America: Problematising the Rental Market for the Chilean Case. *Habitat and Society* 12 (2019).
31 Review the proposals of the company Mi Plusvalía: https://miplusvalia.cl/como-funciona.php
32 PFA is the acronym for Pension Funds Asscociation, or AFP for its Spanish spelling.
33 Information on this Problem can be Found in Central Bank Reports. www.bcentral.cl/documents/20143/1114599/IEF2_2018_rec_evolucion_edeudamiento.pdf/14452d37-0f19-5c52-5ace-7f3e78999798.

5 Real estate profitability

For decades, Chile has fostered the aspiration to be a country where everyone has access to affordable housing. The annual report of the company Socovesa S.A.[1] indicates that by 2019, it planned to build 5,066 new homes, with an investment of UF 9,400,000, expecting sales of UF 15,700,000 (p. 35). *Roughly speaking*, this would indicate that the expected gross return on investment is 67%, with houses to be built at a cost of UF 1,856 per unit and to be marketed at an average of UF 3,100. These numbers are at a company level. Here, we are interested in reviewing how the profitability expectations of these mass housing production companies are passed on to the price and whether, with another production process strategy, these values could be lowered. As we have already mentioned, we are witnessing a housing crisis in Chile today, the clearest representation of which is a deficit of 600,000 units and purchase and rental prices that are increasingly out of line with the purchasing power of households.

Through an empirical analysis of recent investments, the objective of this chapter is to show how much the profitability of real estate businesses influences current prices, how many families are left out of the possibility of buying because of those amounts, and what would happen to values if we adjust the rents that real estate businesses are obtaining to make housing more affordable. For this, we developed an analysis of eight real estate investments of more than ten flats, executed in the last five years and located in different communes of the metropolitan area of Santiago.[2]

These are the results of the evaluation of these cases by communes, for projects between 2015 and 2019 (Table 5.1).

In the evaluation of each project, bank financing of 50% of the total project was considered at an interest rate of 4%, based on the trends in rates over the last five years. The exercise carried out illustrates how the profitability of the projects and, consequently, their final prices are much higher than what is already considered a good investment (around 18% of the Internal Rate of Return). The profitability in the cases studied does not

DOI: 10.4324/9781003348771-5

Table 5.1 Baseline profitability of high-rise real estate projects.

Comuna	*La Cisterna*	*San Joaquín*	*Santiago*	*Cerrillos*	*Independencia*	*La Florida*	*Estación Central*	*Ñuñoa*
Investment (UF)	**673,913**	**853,623**	**641,116**	**441,943**	**777,246**	**1,460,145**	**278,030**	**723,830**
Sales income	1,328,067	2,144,456	2,350,803	1,188,960	2,866,013	3,479,650	1,090,649	2,411,309
Financial costs	*377,391*	*478,029*	*359,025*	*247,488*	*435,258*	*817,681*	*155,697*	*405,345*
Profitability								
Cash flow year 0	−336957	−426812	−320558	−220972	−388623	−730072	−139015	−361915
Cash flow year 1	132,807	214,446	235,080	118,896	286,601	347,965	109,065	241,131
Cash flow year 2	199,210	321,668	352,620	178,344	429,902	521,948	163,597	361,696
Cash flow year 3	485,852	915,867	1,168,997	525,336	1,427,651	1,444,091	553,225	1,162,006
Cash flow year 4	132,807	214,446	235,080	118,896	286,601	347,965	109,065	241,131
IRR	52%	74%	113%	81%	114%	69%	121%	102%

Source: Authors based on data from inciti.com, Servicio de Evaluación Ambiental and CMF.

fall below 40%, with rates of return above 50% in most cases. This profitability represents the entire real estate development process, including land purchase, bank financing and sales.

Given that each project had different flat typologies, in order to compare the cases studied, we estimated a "typical" two-bedroom dwelling for 50 square metres. We obtained that in the cheapest of the cases (Estación Central), the price of a "typical" flat does not fall below 2,800 UF. This requires a dividend of at least $311,076 per month for 30 years and a base down payment of more than 15 million pesos.

Who can afford these prices? To find out, we used household income data from the CASEN 2017 survey in UF, assessing each commune to see which households, categorised by income level, could afford these new properties and who would be excluded. The results of this study are presented in Table 5.2. Since the CASEN 2017 does not offer statistical precision for each commune reviewed, it is important to consider that what we present here is an estimate that would need to be tested against a larger volume of data. In any case, it allows us to illustrate certain relationships between housing prices and income.

It is observed that the prices of the houses offered by the projects studied in each commune have exclusive prices, so that only the higher income segments could opt to buy.

This situation is aggravated when one looks at the time that households in each commune would have to spend to get the footing of the property offered in their municipality. At present, most mortgage loans finance only 80% of the price, so households must have savings of 20% of the value of the house (or find that money in other ways). The prices of the projects studied require households to save a quarter of their monthly income for periods ranging from 4 to 7.5 years.

Table 5.2 Assessment of which socio-economic groups/communes could pay a dividend according to the price of the dwellings we have assessed.

District	*Price of 52 m² flat*	*Initial payment (20%)*	*Mortgage payment (Interest rate 4%)*	*Decile who can pay the price*
Santiago	4,732	26,215,280	500,623	10%
Cerrillos	2,264	12,541,769	239,505	40%
Estación Central	2,940	16,289,615	311,076	20%
Independencia	3,813	21,125,867	403,432	10%
La Cisterna	3,178	17,604,889	336,194	30%
La Florida	3,598	19,933,536	380,663	30%
Ñuñoa	4,120	22,824,800	435,876	20%
San Joaquín	3,224	17,860,960	341,084	20%

Source: Authors' elaboration based on CASEN 2017.

These data indicate that if a three-person household residing in the commune of Santiago, with an income of $770,000 per month, spends 65% of its income on fixed costs of rent, basic food basket and public transport, it will have to save 35% of its income for five years to be able to pay for a foot of a flat in the same commune. The offer of some real estate companies of a down payment in instalments is only useful for high-income segments.

In May 2019, the government announced a package of measures dubbed the Protected Middle Class, which recognises the problem of property prices. The programme included a VAT exemption for homes up to 3,000 UF and an increase in subsidies for buying and renting. These measures are insufficient, as they are unlikely to alleviate the problem of access to credit for low-income households and to absorb the costly down payment (for a 2,000 UF house, the footprint is equivalent to 37 minimum wages). In general, governments have generated palliative solutions that do not constitute robust housing policies on the part of a state that seeks integration in a structural way and not only to reduce the social impacts of some crises that are believed to be temporary. There is an urgent need to try other types of solutions and explore more drastic alternatives to the problem.

For example, if housing were to be considered a basic service (like drinking water or electricity) and because it is a service without substitutes, it is subject to central regulation by the state. In other words, think of housing as a public infrastructure rather than a private consumer good. With this, it would be possible to set a maximum profitability for the sale of real estate projects, in line with other places in the world that have developed radical and effective short-term housing policies in the face of housing crises. In all cases, the role of the state is fundamental. This is despite the fact that these are capitalist societies with a strong free market component. Examples include the Netherlands, Spain, France, Germany, Austria, Australia and some states in the United States, where, with different alternatives, the state influences the housing market in pursuit of the common good.

To advance our idea of regulating profitability, we conducted an analysis based on the same projects evaluated. We set rates of return of between 12% and 18% for these projects. This rate of return remains positive for investment projects anywhere in the world.

If we lower the profitability of the evaluated projects to an internal rate of return of 18% (which is a great deal anyway), the properties would reduce their value by 10% to 54%, making the total dividends to be paid less than 25% of the average household income per commune, which is what is recommended. And here, we want to be emphatic: under this assumption, investors' returns are still very high.

Real estate market agents have for years insisted on the stubborn hypothesis that the higher the volume of housing production, the lower the price, following the positive premise of the supply/demand rule. In their argumentation, they rarely mention the role that expectations of profitability of housing projects have played in shaping these prices. While there are well-known and measurable costs such as construction, permits, taxes and even land value, little is said about profitability, and records are hard to find. However, data can be collected for different projects and a line of income and expenditure can be built up from information that some companies share with public bodies. They can be explored in companies' annual reports and in some environmental impact statements from the Environmental Assessment Service, but the figures given do not match either. In our case, we have resorted to archival work to uncover the profits of some projects which, moreover, seem to us to have contributed little to the urban environment in which they are located. Depending on the project, land price and profitability have similar weights on sales revenues. It is important to illustrate to people who are not familiar with project evaluation how each case should be investigated in order to reach more accurate conclusions about the cost and profit structure of real estate projects.

The following evaluation seeks to illustrate the cost structure of a project and thereby show how the profitability of projects in the same commune has varied, but ten years apart. Specifically, an example is developed based on a financial evaluation model for real estate projects developed by the financial expert Dr. Julio Aznares, who has published this method with many methodological details that have allowed us to recompose the series with a different approach to the one we have developed for our series of analysis. The methodology is available in the book edited by José Miguel Simian and Veronica Nikiltschek, from the Universidad de Los Andes, which was sponsored by real estate companies for its publication,[3] so we understand that the model is validated by the industry itself. These are the results of the update of values and yields (Table 5.3).

The net variation in NPV profitability is 5,325% between a project evaluated with 2009 values and another with 2019 values. If we assume the expenses attributable to bureaucracy, reimbursable contributions resulting from mitigation, transit and environmental studies that could be part of this cost structure, the maximum value to be paid for them, ensuring a profitability of 12% + CPI, is almost 700,000 UF. This profitability is influenced by the fact that the speed of sale for the project is much higher than in 2009, so that a project does not take more than 36 months from start to sale in the commune of central Santiago. So, in light of the results, it cannot be said that the price increase has nothing to do with the total profitability expected by investors, which is evident in projects executed by real estate companies with a national presence.

Table 5.3 Profit assessment updated to 2019 values measured in UF.

Cash income	*2017*	*2018*	*2019*	*Total 2019*	*Reference 2009*
Totals		803334	222119	1036557	**443778**
Expenditures					
Land	−78387			−78387	−25328
Seller of land	−995			−995	−520
Construction costs	−48970	−143670	−93327	−285967	−255836
Architecture and engineering	−11442	−8175		−19618	−10249
Landscape	−1994	−1711		−3706	−1936
Technical inspection	−2144	−2144	−2144	−6431	−3360
Municipal rights	−2938			−2938	−1535
Services	−7124	−2955	−530	−10610	−5543
Sales costs	−14038	−12981	−7061	−34080	−20912
Administration costs	−18458	−8724	−5178	−32360	−16906
Legal costs	−4402	−3116	−1648	−9167	−4789
Taxes	13075	−49780	−13355	−50059	−26153
Total	−177818	−233257	−123243	−534318	**−373067**
Cash flow					
Total	−177818	570077	98876	251858	**70711**
Cash flow for banks					
Total	−10534	−181187	−89977	−268521	**11577**
Cash flow for investors					
Total	−167284	751264	188853	520379	**59134**
IRR project flow	237.10%				10.20%
IRR investor flow	373.00%				20.10%
NPV	UF 689903				UF12655

There is a defence of the excessive increase in housing prices in Chile that emerges from the spheres of real estate power. From these spheres of power, the evidence to be shown is chosen, with which the scope of knowledge is managed in defence of particular interests, which privilege some results over others. This produces a series of technocratically validated decisions. From science, one can have discussions about methodological approaches to study certain phenomena, about sampling or interpretations of the latter. Science-based methodology allows for repetition, testing and, above all, doubt (which for the authors of this book is fundamental, since it generates discussion). We understand that doubt must be reduced or eliminated in the business sphere, as it generates uncertainty, which jeopardises investment. But the results are compelling and real estate projects are tremendously profitable at the expense of people, whose access to housing is increasingly difficult. However, the reason large real estate companies behave abusively is because they have a lot of pockets to fill. This is discussed in the next chapter.

Notes

1 Socovesa S. A. Annual Report 2018. *CMF Chile*. Santiago, 2018. www.cmfchile.cl/institucional/mercados/entidad.php?mercado=V&rut=94840000&grupo=&tipoentidad=RVEMI&row=AAAwy2ACTAAABy3AAF&vig=VI&control=svs&pestania=49.

2 The data were obtained from inciti.com and the Environmental Assessment System. The methodology is available in: Vergara-Perucich, J. F. and Aguirre-Nuñez, C. Housing Prices in Unregulated Markets: Study on Verticalised Dwellings in Santiago de Chile, *Buildings* 10 (2020): 6.

3 Simian, José Miguel Simian and Verónica Niklitschek, *La Industria Inmobiliaria En Chile. Evolución, Desafíos y Mejores Prácticas*. (Lima: Pearson—ESE Business School Universidad de Los Andes, 2017).

6 Housing and financialisation

Financial power is a system of networks of capital flows controlled by agents who ensure their safety and who understand how to direct those flows to areas of society where it is most profitable to invest and where there is the greatest certainty that capital will grow. The financial world is complex because it has been designed to be complex, cryptic, and difficult to read. This logic seeks to generate a barrier to entry constructed by financial agents, who hold the keys to the door, thus ensuring that they are indispensable. However, their objectives are less complex given that financial power is only driven by income and security. It is not difficult to understand that, in this quest, states do much to generate security for financial investments and thus get a head start in a global competition for financial capital to invest in local companies. The rent is offered, in the last link, by consumers who prefer one or the other company, but this rent can also be oriented towards capturing social areas of high demand. In concrete terms, when financial power is installed over basic goods such as water, electricity, gas, pensions, health, education or housing, demand is assured because they are basic necessities and it is impossible to live without them. Therefore, financial investment in these assets generates constant and secure rents. In Chile, housing undergoes a process of financialisation.

Raquel Rolnik warns about the social danger of the housing system relying on the financial market. Rolnik speaks of the financialisation of housing to characterise a process in which the speed with which people's indebtedness grows, in order to gain access to their own home, turning property into a financial asset.[1] As we have already explained, debt has been installed in Chilean households as a mechanism to alleviate the high costs of living and to facilitate participation in what Gabriel Salazar has called the society of hyper-consumerism.[2] The ease with which households have access to credit instruments such as credit cards and bank loans is highly convenient for banking and financial institutions in general, taking advantage of the fact that Chileans are also good payers in relation to the regional context, so

DOI: 10.4324/9781003348771-6

much so that they are able to borrow a second and even a third time to pay off an initial debt that they would not otherwise be able to repay. Much of the housing that is marketed in Chile, particularly new housing, is acquired through mortgage loans, so there is an ostensible dependence between debt and housing, perfectly fitting with Rolnik's descriptions of the processes of housing financialisation and, consequently, also adopting its dangers. However, Manuel Aalbers argues that financialisation is not necessarily just indebtedness but also the incursion of financial actors into the process of housing production and acquisition, thereby changing the focus of housing for living, turning it into an asset for rental purposes.[3] As Aalbers himself states, financialisation is a broad interpretative framework for a specific problem, which is the importance that the financial market now has on a basic human right such as the right to habitat.

In Chile, the financialisation process accelerated in 2001, after the capital market reform. At that time, new investment instruments were created and real estate development took a leading role.[4] Among other factors, the powers of financial institutions (including PFAs) to invest in the real estate market were expanded. Also, some larger real estate companies began to go public for share purchases and, with this, the partners of real estate companies diversified, thus changing their internal power structures and decisions to became more oriented towards the profitability of real estate investments. Thus, banks, investment companies, pension funds, stockbrokers and insurance companies become shareholders or partners of national real estate companies. In doing so, the *ethos of* these companies shifts from housing production to income production.

The financialisation of housing is a socially complex phenomenon that already caused a planetary crisis in 2008. For David Harvey, it is urgent for housing to move out of the markets and into the public domain, like beaches, parks or streets,[5] but in the particular case of Chile, this is a complicated approach to fulfil because the housing market has begun to be filled by actors seeking rent, who come from different worlds: professionals under 50 with average or relatively high incomes who buy several properties to increase their income by renting; people who, knowing that their retirement will be miserable, seek protection in the creation of real estate assets in case of emergency, illness or old age; the first professionals of a family who buy several homes to have a place to house their families. But there are also the investment groups who buy properties in bulk just to rent them out. Now, what will happen if the whole society invests in several houses to ensure a better quality of life? Who are they going to rent or sell to if everyone is already a homeowner? This last rhetorical question is answered by reviewing the causes of the economic crisis in the United States in 2008, seeing how the real estate market did not understand its own risks and how the banks did not make a prospective analysis of the real dangers represented by a structural failure of the mortgage system, built without safeguards, seeking easy, short-term and sustained income over time. This is the justification for Harvey's call to remove housing from these speculative markets and place it in the public domain. In Chile, the current constitution makes this task difficult because it assigns a subsidiary role to the state, does not enshrine the right to housing and does not allow the state to act in competition with private actors, among other limitations.

To move forward, it should be understood that for all nations, housing should be considered a national good for public use, but with a specific family allocation, where the quality, delivery and organisation of residential units should not depend on people's ability to pay or on the interests of the banks, but on greater goals such as social integration, reducing spatial segregation, ensuring a dignified space in which to develop a life project and contributing to the general wellbeing of the population. Of course, as Harvey, Rolnik and other researchers argue, this is very difficult to do within the margins of neoliberalism, but given the Chilean reality since October 2019 and the constitutional process that has emerged since those years, possibilities are opening up to rethink the social role of property, with special emphasis on housing.

It is important to remember that until 1973, Chile moved steadily towards a system of urban development in which the state played a leading role in organising the city, coordinating the interests of civil society and the private sector in order to urbanise while protecting the common good.[6] This was mainly done by planning the supply of housing through public urban management offices, such as the Housing Corporation (CORVI) and the Urban Improvement Corporation (CORMU). There was also a private housing market but under strong regulation and in direct collaboration with the authority, which had the final say.

For this to work, private property was subordinated to its social role in its period of greatest splendour, between 1967 and 1973. This followed the constitutional reform instigated by Eduardo Frei Montalva, which expanded the state's powers to shape territories by pursuing social objectives over private interests.[7] The logic was also economic: the precariousness of working-class housing, commuting times and the habitat itself had a direct impact on productivity. In addition, the under-exploitation of the city in relation to its centres was inefficient, as was the under-exploitation of the agricultural land that was distributed among peasants with Frei's own agrarian reform. In no way did this law prevent the existence of private property, but rather subordinated it to the national interest. It was certainly a different Chile, with a GDP far from that of developed countries and with precariousness in several productive sectors. In that Chile, there were not as many public resources as there are today, but there was an awareness of state planning aimed at the common good as a measure of social welfare and also as a productive strategy. Despite being a poorer country, public and private actors agreed to move towards urban development with an emphasis on quality.[8] This can be measured economically, given that the social or public-private complexes built at that time have not ceased to gain in value since they were handed over after almost 50 years. In this sense, architectural works of great quality for the time stand out, such as the Remodelación San Borja, Remodelación República, Villa San Luis, Conjunto Atahualpa Yupangui, Unidad Vecinal Portales and Unidad Vecinal Providencia, among other examples of housing planned by the State in a predominant role, in well-located complexes, with a high standard of design and material quality. These characteristics made it competitive in comparison with what the private market was producing at the time. Moreover, the social housing of those years when Chile was a very poor country is still far superior to the social housing produced in today's Chile, an OECD country on the verge of making the leap to development and with a GDP per capita similar to that of some European nations. Before neoliberalism, being a poor country was no excuse for a non-rentier approach to housing. Neoliberalism was an ideological tsunami that swept away all ideas of the public, starting with the state. Housing was part of that wave of institutionalised individualism.

In terms of housing, the advances in housing began to be dismantled with the tax exemptions for construction companies in 1976 and the National Urban Development Policy of 1979. Following these reforms to the housing production system, the theory of supply and demand (in hypothetical perfect competition) took over the planning and management of territories. The state was no longer needed to plan the city, but everyone on their private property could do what they wanted as long as they adhered to the communal regulatory plans (perhaps the only and limited form of planning in Chilean cities). Consequently, private profitability became the main driving force behind the development of cities, overriding the common good and social problems. Thus, the world saw the birth in Chile of what is known as "neoliberal urbanism".[9]

Urban neoliberalisation changed the way cities were ordered and ended up encouraging segregation,[10] increasing the value of land[11] and perpetuating the production of neighbourhoods for the rich, the middle class and the poor.[12] The cities were built from the private estates outwards. The resulting urbanism was composed of the aggregation of buildings arranged on the territory, with no order other than knowing where to focus investments that ensure returns in the short term. It is bold to say that neoliberal urbanism in Chile generated an urban form, given that there was no objective image of the city, much less a collective idea of how to live in the community.

Neoliberal urbanism produced rents and privatisations that gave life to a space of difficult form resulting from the imposition of a monetarist vision on social construction. Thus, the Chilean city was organised in order to sell it and not to inhabit it.

From the beginning, urban planners warned of these effects, but they were always ignored by the authorities. In 1980, Francisco Donoso and Francisco Sabatini already warned that these measures would accelerate urban segregation, something that economist Pablo Trivelli would support. Not much progress has been made since then, in terms of the guiding principles of Chilean urban planning. Despite some palliative measures of public investment (capitalised by the private sector, of course), the model remains the same. To this day, the structure of urban development follows the same logics of investment profitability and the role of the state remains subsidiary, acting more as a facilitator than as a regulator, much less as an actor that organises urban form. In other words, it operates neoliberally.

Unsurprisingly, this logic treats housing like any other product for sale and denies its particular characteristics. First, each property has a unique and unrepeatable location in space and therefore, it is key to understand that there is a monopoly of location. Second, housing competes in a market typology where attributes help to compensate for a less advantageous location (e.g., living in a detached house with a courtyard, but far from the centre). Finally, housing is a good with no substitute in a captive market: one cannot live without housing and therefore, it should not be seen as any other market good. Leaving its prices to the free market is irresponsible, and even more so if the ability to pay in a country is as unequal as in Chile. Both concentration and price should be framed in legislation that seeks to ensure equal access to a secure property and that it does not become an exclusive monthly expense. As George Monbiot has suggested,[13] housing should cease to be the main burden on our wages, which are also very low. In many parts of the world, housing is a human right, and Chile has even signed international agreements that recognise housing as such. But if it rests on the financial market, as Rolnik explains, it is unlikely to be equally distributed to the population as a whole. In a neoliberal society such as ours, money is king, and with brutal levels of inequality, purchasing power ends up defining the standard of living, the level of spatial dignity, and access to the city.

In this sense, it is essential to not lose sight of the problem related to purchasing power in relation to access to housing in Chile. As this is a captive market with no product valuation regulations, the price adjustment will be made towards the greatest possible ability of consumers to pay. Given the growing concentration of property ownership for rental income,[14] while the price of housing continues to rise, the dream of home ownership for the less well-off social classes is beginning to crumble. The data corroborate this analysis: while wages increased by 20% between 2010 and 2017, house prices increased by 53%.[15] It is possible that this increase is related to the delinquency of Chileans, which grew by 10%, reaching 4.6 million debtors in arrears with an average of 1.7 million pesos per person,[16] as we mentioned previously. For all of the above reasons, we maintain that it is essential that states establish policies to regulate the price of property and ideally do so by participating directly in the fair allocation of property, based on universal mechanisms that allow for a reduction in segregation. It is also desirable for the State to be able to separate housing from the financial world, for which it must assume a leading role in the productive processes and in the management of supply.

As stated above, the housing market can be highly lucrative and at least it is managed in scenarios where profitability expectations have been progressively exceeded. We have already shown how difficult it is to get the information from research to obtain real data on projects in order to know their expected returns. Now, just as difficult as this data is to know how extensive the network of actors who have vested interests in the housing market is and who, by action or convenient omission, benefit from the fact that the price of housing continues to rise. What is happening in the spheres of power in the real estate world? We have looked into the public reports of the big real estate companies, the ones that are all over the country and that are accountable to the financial market commission (CMF). These are the ones that lead the housing financialisation process. Reviewing the annual reports and annual reports of the six main real estate companies in Chile, we can see that during the 2018 financial year, net profits reached 116,414 million pesos. Within those profits, at least 3,023 million pesos went to banking entities that appear as shareholder partners of these main real estate companies. This shows a strong sign of financialisation, where banks participate in the housing market not only by lending money to companies to build and to consumers to buy, but also by improving their own direct liquidity by owning the main companies in the country. If we add investment funds, mutual funds and rentier companies that have shares in real estate companies, we reach a sum of 26,865 million pesos of net profits from real estate activities, whose main line of business is not construction or architecture, but renting. This is a high and worrying factor, given that housing would be, as Vicente Domínguez said, fully fulfilling its lucrative spirit in exchange for receiving important volumes of liquidity from rentier entities with financial objectives.

Table 6.1 Aggregate 2018 net profits registered with the Financial Marking Commission on partner banks of major real estate companies.

Financial institutions	*Income in Chilean Pesos*
Banco de Chile	1025590420
Banco Itau	411237334
Banco Santander	1358233957
Banco de Crédito e Inversiones	228348880

The presence of financial institutions in the process of buying, selling and defining investment objectives is a complex situation (Table 6.1), given that they are positioned at different stages of a production chain. They extract value mainly from consumers, to whom they will then lend money to make up for financial shortfalls that, among other things, could be caused by high housing values, benefiting the banks themselves in a constant *loop*. This financialisation of housing[17] has already generated harmful effects in the world before and it has not yet finished doing so in Chile, so it seems to be the time to take action on the case.

In order to provide more background on the existence of financialisation in Chile, a brief analysis of the explanatory factors that work best statistically for housing prices at present is applied. For this purpose, an exploratory study with a quantitative approach is applied using the econometric technique of ordinary least squares analysis for different observations in a linear stationary trend to review the influence of financial factors on house prices, in relation to other factors identified as influential in the literature. A hedonic price model is applied to the natural logarithm of house prices as the dependent variable, in relation to other independent variables that the literature on house prices or financialisation identifies as fundamental. The semi-logarithmic formula is developed as follows:

$$\text{Ln(house price)} = \alpha + \beta 1 * X1 + \beta 2 * X2 \ldots \beta i * Xi + \varepsilon$$

Where α is the constant intercept of the regression curve, εi is the random error and βi are the estimated coefficients for each dependent variable Xi, which are listed below:

Variable	*Mean*	*Standard deviation*	*Source*
Mortgages granted (N)	4469.84	449.19	Financial Market Commission
Stock market value of Real Estate Companies (N)	410.55	158.57	Investing.com
Land value (UF/sqm)	27.15	40.67	Inciti.com

Variable	*Mean*	*Standard deviation*	*Source*
Housing sales (N)	4226.96	1031.03	Chilean Chamber of Construction
Wage Index	5.45	0.99	Central Bank of Chile
Construction costs index	4475.23	182.6	Chilean Chamber of Construction
Woman purchasing housing	0.33	0.07	Inciti.com

A semi-logarithmic estimation is developed for the dependent variable under study on house prices. The data record monthly variables for a ten-year period between 2009 and 2019. These variables are difference-converted to ensure seasonality.

It is important to mention that in Chile, there is no direct access to the price of housing transactions, but that these records are consolidated by the Real Estate Registries (Conservadores de Bienes Raíces). Access to their books is public, but it requires a lot of time and dedication. In this case, the database for house prices was acquired from the company www.inciti.com. The time of analysis considers ten years. It is important to mention that Law 20.780 introduced a VAT charge on the sale of homes whose construction costs were higher than 2000 UF. This is applicable to regular home sellers. This tax came into effect on 1 January 2016. However, in the data observed for this analysis, the average monthly increase in the price of housing 28 months before that date was 0.631%, while the average for 28 months after that date was 0.504%, for a total average variation of 0.59% for the period analysed. So if there is an influence, it does not break with the average variance of the monthly series, which is 0.033%. This lack of variance is also explained by the fact that the dataset contains second-hand dwellings and that the average size of the flats decreases between 2016 and 2019, so even if the VAT tax bracket changes, the taxed dwellings were not all of them. Even so, the analysis technique is not by time series, which reduces the impact of this type of event. In order to reduce the noise in the sample, a smoothing technique is applied by averaging over the four closest values of each observation.

The semi-logarithmic regression applied is based on a stepwise method that allows to check if an independent variable offers statistical significance to predict the variance of the dependent variable, excluding the independent variables with low capacity from the final model to predict the dependent variable and leaving only those with the best fit. This regression technique

Table 6.2 Results of ordinary least squares analysis. Adjusted r2 factor = 0.41, model p-value less than 0.05. Flags indicate the significant variables.

Variables	*Coef.*	*Std. Deviation*	*t*	*p-value*	*Flags*
Const	7.9354	0.0129658	612	1.44E-194	***
Mortgages Granted	−0.00036005	6.91E-05	−5.211	8.97E-07	***
Stock Value Real estate companies	0.00463015	0.00099883	4.636	9.93E-06	***
Land Value	−1.50E−07	1.32E-07	−1.14	0.257	
Housing Sales	−0.00012633	5.58E-05	−2.265	0.0255	**
Wages	0.046295	0.0799376	0.5791	0.5637	
Construction Costs	−0.00029444	0.00048379	−0.6086	0.5441	
Women Purchasing Housing	-9.0478	1.63975	−5.518	2.4E-07	***

is especially useful for exploratory studies. The results are summarised in Table 6.2.

The results indicate that the most significant explanatory factors for house prices are the delivery of mortgage loans, the value of shares of real estate companies on the stock exchange and female homebuyers. Of these factors, the price of shares on the stock exchange explains the increases in the price of housing, while the delivery of mortgage loans and women buying houses explain the decreases in the price of housing. House sales also appear as an explanatory factor; the more house sales increase, the more house prices fall. Land prices, wages and construction costs were not statistically significant for this model. In short, for this explanatory test, it is observed that the price of housing is affected by two key financial factors: the availability of mortgage loans and the stock market performance of the shares of real estate companies. In other words, empirically, it can be said that the financialisation of housing in Chile has a real impact on the price, which merits an evaluation within the frameworks of political economy to review how this impacts inequality.

Security of housing tenure is one of the main priorities of Chileans,[18] an approach sustained by a history of housing policies aimed at facilitating access to home ownership for Chileans.[19] The state has neglected its role in relation to housing markets regulation, focusing only on subsidiarising areas the real estate market is not interested to tackle, such as economic housing. Currently, it is estimated that 63.7% of households are in the process of acquiring or owning a home, and 20% are renters. Given the recent evidence and the experience of the 2008 real estate crisis, a broad domain of private home ownership should take certain precautions.[20] In the Chilean case, much of this access to property is developed through indebtedness to financial institutions, as we discussed previously. This special situation poses the question of how financing the access to housing means opportunities for some and barriers for others.

In 2002, Jean Cummings and Denis Di Pasquale along with Edward L Glaeser and John R. Meyer, in *Chile Political Economy of Urban Development*, argued that housing policy in Chile contributes to spatial distortions that, among other consequences, limit access to the formal rental market in which wealthier households can participate more freely. In turn, the concentration of investment and the structuring of the new housing business in real estate projects benefit from the processes of the high demand for housing, a constant situation in Chile in the face of the financialisation of an important part of the population. This is under a phenomenon in which the price of housing is fixed by the scarce need to increase the competitiveness of the market,[21] whose main goal is rent and not the provision of housing. According to recent results, households in Chile spend 56% of their monthly income on housing rent, public transport, and daily food, but when segmented by socio-economic deciles, it is observed that while households in the lowest income quintile should spend 118% of their income on fixed expenses, that is, formal urban living is financially unsustainable, households in the highest income quintile cover their basic needs with 18% of their monthly household income.[22] Additionally, between 2009 and 2015, the number of households in the highest income decile in Chile that receive income from renting urban property increased by 22.98%.[23] In the specific case of Chile, there are few studies on the housing rental market, and with this, there is little information on how these processes affect the quality of life of households from an economic perspective. We do believe this is a matter of political economy of housing.

In 2018, a new concept for analysing rental markets appeared. Kath Hulse and Margaret Reynolds introduced the concept of investification "to explain a process whereby disproportionately high levels of household investor purchases in disadvantaged suburbs contribute to higher prices/rents and to the persistence of socio-economic disadvantage, as properties are rented on the private market to low socio-economic households, indicating replacement rather than displacement".[24] In their approach, investification is presented as a useful framework to analyse the processes of gentrification and financialisation in cities with a focus on letting markets. This chapter explores how applicable this framework is in the case of Santiago de Chile to adapt the method employed by Hulse and Reynolds for the particular context of Chilean cities and contribute to our discussion on the political economy of housing. The chapter exposes how investification occurs in Santiago and why it is quite relevant to understand it as an obstacle to advance towards more just cities.

In the neoliberal context of urbanisation processes, the success of a city is measured by the profitability of its urban investments. In these processes, the role of financial capital is fundamental,[25] which in terms of housing is presented as a safe activity for wealthy households, but risky for those who depend on debt to access security of tenure.[26] This has led to a global crisis of tenure insecurity.[27] In this crisis, tenants are highly exposed to the interests of landlords whose rent-seeking tends to generate patterns of exclusion with purchasing power. During the COVID-19 pandemic, tenure insecurity was correlated to a major risk of contagion.[28] In the case of Santiago de Chile, the relationship between housing quality and COVID-19 infection has been significantly high.[29] When social rules are configured through laws and norms to provide on the basis of deregulated free competition, as is the case in Chile,[30] individualism prevails and with it, the purchasing power of individuals ends up defining the socio-economic configuration of neighbourhoods, underpinning a process of financialisation where indebtedness influences the creation of modes of tenure focused on increasing the exchange value of housing rather than improving its use value.[31] This affects feelings of social security in everyday life and access to housing.[32] In this context, access to private property is one of the main aspects of generating security in the face of a daily life threatened by risky neoliberal financial scenography.

The process of exclusion of neighbourhood residents by urban renewal processes and the replacement of residents with higher purchasing power than previous residents has been theorised as gentrification.[33] We understand the investification as this variant, which categorises an urban process in which an exogenous investor buys housing in lower-income neighbourhoods, puts pressure on the price of such properties and rents these properties for lower-income tenants. The investment occurs in urban areas where such investors would not live. Investification is a model of rent extraction from the very need for housing of social classes with less purchasing power than the investors. This business model may well be used to install a constant process of rental income flow, especially when there is a housing deficit. This phenomenon in Chile has been called cash cows: properties that are leased without vacancy and increase their value as assets over time.[34] In this chapter, we will shed light on this theme for the case of Santiago.

The results here presented were deductively elaborated, aiming to test the theoretical basis of investification for the case of Santiago de Chile. As such, the approach is exploratory and employs quantitative techniques of analysis framed under the scope of urban economics and demographics. Based on the definitions of Hulse and Reynolds, it is assumed that an investified property is one that belongs to a property owner who has a portfolio of investment properties. For this study, it has been established that owners of three or more dwellings are classified as investificators. From this premise, then, datasets are constructed to determine those urban areas with higher levels of investification. Categories of low, medium and high investification are created with reference to the geographic units composed of census tracts of the year 2017 for Greater Santiago. The objective is to take these areas to characterise them and propose a profile of urban areas where investments in housing for rent in Greater Santiago are located.

There are three different sources that are brought together to form the databank to be analysed. The first consists of a survey of transactions registered in the Santiago Real Estate Registry for new housing, filtering the buyers who purchased three or more homes between 2009 and 2019. For this dataset, we have the typology of the dwelling, the sales values, and the value per surface area. The source of this data was obtained through the Inciti.com website. The second databank corresponds to the rental properties available in the Metropolitan Region during 2019. This data is obtained from the information available on www.portalinmobiliario.com, a site for sharing properties for sale and rent. This information is key to testing two critical issues: to identify whether residents in each census tract have the purchasing power to pay the rental value and to explore how advantageous it is to buy to rent in each census tract. From the first dataset, 49,914 housing transactions are identified that are categorised as potential investificated purchases. Of these, 38,542 have been geo-referenced for this analysis in 9,987 unique addresses. Then, a spatial joint is run in the QGIS 3.22 software, between the base of investified transaction records and rental values to consolidate a total sample of 27,311 cases of properties whose owners have two or more dwellings whose letting price is known. These cases are in 1,213 census tracts out of a total of 1,885 for the Metropolitan Area of Santiago. A summary of these records by district can be found in Table 6.3. Also, for each census tract, we calculated the average household income according to educational level by commune, which was obtained from the CASEN 2017 survey weighing the sample by the communal expansion factor.

Table 6.3 Summary of records of cases of investification by communes in Greater Santiago.

District	*Average Income of Main Job per Household (USD)*	*Investificated Cases*	*Average Sale Prices of Properties analysed (USD)*	*Average Sale Prices of Properties analysed (USD)*
Santiago	1167	8747	48096	48100
Independencia	691	3683	54239	54255
Estación Central	703	2131	20653	20671
San Miguel	1231	1564	68183	68177
Ñuñoa	1571	1477	23199	23206
Quinta Normal	722	1402	61999	61986
Macul	862	1227	69800	69794
Providencia	2528	1141	158999	158980
La Florida	925	1061	69032	69031
Recoleta	1054	659	70406	70426
La Cisterna	807	423	45954	45971
Peñalolén	1416	353	32738	32742
La Reina	1757	321	229041	229045
Huechuraba	878	317	80833	80827
Puente Alto	806	317	37507	37499
Las Condes	2618	307	242702	242683
San Bernardo	696	291	34799	34810
Pudahuel	695	197	37507	37507
Cerro Navia	562	196	25018	25002
Vitacura	2737	178	340187	340196
Renca	755	171	39204	39215
San Joaquín	525	163	35243	35240
El Bosque	634	156	34395	34382
Lo Prado	528	137	39932	39939
Conchalí	649	125	21785	21782
Lo Barnechea	3104	122	408451	408437
Maipú	921	100	9740	9756
Quilicura	836	80	38517	38506
Cerrillos	653	73	57230	57250
Pedro Aguirre Cerda	649	57	37143	37127
La Granja	483	48	24897	24897
La Pintana	473	43	19925	19944
SAN Ramón	617	37	22068	22081
Lo Espejo	565	7	4405	4386
Overall	1053	27311	74818	74819

The third data collection process consisted of mapping the location of different urban services and functions by census tract to review those amenities and attributes that help to explain urban planning criteria for the location of buy-to-let housing investments. This data was obtained from the sample constructed by Correa, Aguirre and Vergara, who facilitated the dataset.[35]

This way, the resulting model will make it possible to define in which sectors new investification processes could be initiated from ongoing urban projects, transformations or renovations, summarising these attributes for each census tract and each commune. In order to unify the samples for the statistical study, a gap index has been created for the variables referring to urban services and functions, which are indicative of greater or lesser relative access to amenities and attributes for each area of analysis. The base variable is composed of the metric of each series that weighs the higher values for each amenity or urban attribute. The formula applied for each index is explained below:

$$U = -1 \times \left(\frac{(Max\,X - X_i)}{Max\,X} \right)$$

where U is the resulting index factor, Max X is the maximum value of the variables under analysis for the census tracts and Xi is the specific value for each census tract. The closer the value is to 0, the greater the accessibility to amenities per census tract.

With geo-referenced and cartographically represented data, different initial visual explorations are carried out to identify those areas of the city where there is a greater propensity for investification. Having the data geo-referenced facilitates the task of identifying the proximities and specific characteristics of the investificated census tracts under study. From these explorations, it was decided to generate four levels of investification: (i) no investification (zero); (ii) low investification (from one to 34 cases); (iii) medium investification (35 to 72 cases); and (iv) high investification (from 73 to 1033 cases), based on sections of cases equally distributed in the total sample.

To mathematically determine the urban areas where there is greater relative dependence between investificated properties and urban characteristics, a Pearson correlation study is carried out between the concentration of investification cases and the other variables. The formula for calculating the Pearson correlation is as follows:

$$\rho_{x-y} = \frac{CovXY}{\sigma x \, \sigma y}$$

where CovXY is the covariance of (X, Y) σx, is the standard deviation of the variable X and σy is the standard deviation of the variable Y. To interpret the results, the closer the correlation factor is to zero, the lower the dependence of the variables. If the result is less than zero, an inverse relationship is established (i.e., if one increases, the other decreases). Conversely, if the correlation is greater than zero, this is a positive correlation and there is a direct relationship between the variables. Finally, we check whether the rental price is significantly different between the medium and high investification areas in relation to the rest of the census tracts with low or no investification. For this review, an exploration between categories and rental value per m^2 is applied, reviewing the descriptive statistical results using a box-plot diagram to represent the results.

As result, we can build a brief profile of where investificators' investments are made, specifically referring to natural persons. Companies have been excluded, as we sought to identify the specific role of individuals who may live in the units purchased. There are 11,703 persons with three or more property acquisitions in the last few years between 2009 and 2019. These persons have acquired an aggregate total of 49,969 dwellings, which average a value of USD 109,287 and an average surface area of 67 m^2. The districts with the highest concentration of areas with a high level of investment are Santiago, Independencia, San Miguel, Macul, Quinta Normal, Ñuñoa and Estación Central, which are mostly peri-central communes, with good access to public transport and with relevant structuring avenues. At the district level, the average rental price is variable among these communes, as well as the dominant socio-economic groups, which can be reviewed in Table 6.4.

Of the total number of properties registered, 27,311 cases of properties whose owners had three or more properties were geo-referenced. According to the records, the concentration of investificated cases tends to be located on a north-south axis of the city composed of Avenida Independencia, San Diego and Gran Avenida José Miguel Carrera. Metro Lines 2 and 3 also cross this axis as structural transport references. Another significant group of investified areas is located around peri-central districts, within the ring road of Avenida Américo Vespucio (Figure 6.1). Although there are cases of investification in 64.35% of the census tracts in Greater Santiago, the high and medium concentrations are mainly associated with structural axes of the city and areas of real estate development that have been triggered by the arrival of metro stations in the last 20 years, in addition to specific characteristics of each sector that will be analysed below. The census tracts with a high and medium concentration of investigated cases correspond to 7.85% of the total number of census tracts in Greater Santiago, which shows an important concentration of investment in specific urban areas.

Table 6.4 Categories of investificators by communes, indicating average rental price and dominant socio-economic group in cases studied.

Districts	*High Investification Zones*	*Mid Investification Zones*	*Low Investification Zones*	*Census tracts investificated*	*Average Letting Prices (USD/m²)*	*Main Socioeconomic Level*
Santiago	29	28	49	106	11,24	ABC1
Independencia	10	2	13	25	9,46	C3
San Miguel	9	4	11	24	8,91	ABC1
Macul	5	3	24	32	9,49	C3
Quinta normal	5	4	19	28	7,82	D
Ñuñoa	4	10	28	42	10,54	ABC1
Estación Central	4	0	32	36	6,30	D
Recoleta	2	3	37	42	5,18	D
La Florida	2	6	92	100	5,07	D
Providencia	1	6	43	50	11,44	ABC1
La Cisterna	1	2	18	21	9,02	C3
San Joaquín	1	0	21	22	7,24	D
Huechuraba	1	2	16	19	5,67	D
Vitacura	0	0	20	20	13,24	ABC1
Lo Barnechea	0	0	18	18	13,20	ABC1
Las Condes	0	0	59	59	12,93	ABC1
La Reina	0	1	26	27	7,31	ABC1
Quilicura	0	0	25	25	4,81	C3
Peñalolén	0	2	35	37	4,45	D
Cerrillos	0	0	18	18	3,51	D
Conchalí	0	0	23	23	3,51	D
Maipú	0	0	25	25	3,42	C3
Lo Prado	0	0	29	29	3,25	D
Pudahuel	0	0	47	47	2,61	D
El Bosque	0	0	35	35	2,36	D
Puente Alto	0	0	103	103	2,22	C3
La Granja	0	0	20	20	1,73	D
La Pintana	0	0	20	20	1,28	D
Renca	0	0	37	37	1,18	D
Pedro Aguirre Cerda	0	0	20	20	0,75	D
San Ramón	0	0	15	15	0,71	D
Cerro Navia	0	1	24	25	0,00	D
Lo Espejo	0	0	6	6	0,00	D

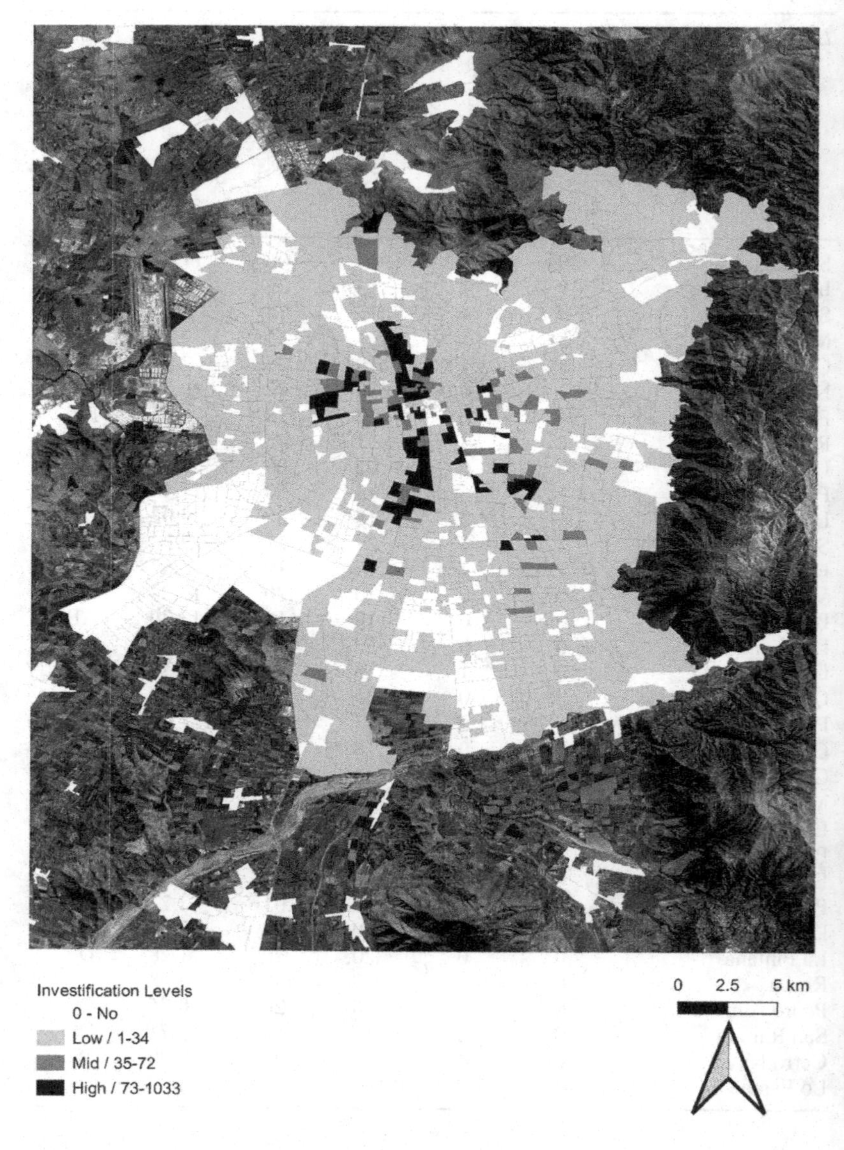

Figure 6.1 Classification by levels of investification according to census tract in Greater Santiago.

Highly investified zones are concentrated in communes within the Américo Vespucio ring road, with a special presence in the north-south axis composed of Independencia, Manuel Rodriguez, San Diego, Gran Avenida and Vicuña Mackenna streets, with different intersection axes in Alameda Libertador Bernardo O'Higgins, Santa Isabel and Avenida Departamental. For the most part, the high investification areas are made up of socio-economic groups ABC1 or C3, 73% of which are working-age households. One of the most striking aspects of this characterisation is the high presence of homes with *allegados*, who are individuals or families living as permanent guests in a friend or relative's home without paying rent in an informal kinship care model, averaging 27% in the high investification census zones, which also have a rental price of housing that represents an average of 61% of the income of people located in the same census zones under study (Table 6.5). The latter is critical, as it would indicate that such housing is not affordable for the current residents of each neighbourhood, which could be indicative of an ongoing gentrification process. Another key aspect is the high presence of non-national immigrants in these census tracts, who represent 8% of the resident population, a value that for Chile is high, considering that the immigrant population registered in the last 2017 Census is 4%.

Table 6.5 Summary table of general characterisation of census tracts with high investification cases.

District	*Zones (N)*	*Cases in Tract*	*Average Income per Zone (USD)*	*Letting Price (USD)*	*Letting price/ Income Ratio*	*Main S.E.L.*	*Informal kinship care (Allegados)*	*Inhabitants*	*Working Age Population*	*Immigrants*
Santiago	29	641	1270	673	0,53	ABC1	0,33	104.316,00	0,82	0,1
Independencia	10	233	797	510	0,64	C3	0,29	49.138,00	0,75	0,01
San Miguel	9	223	1453	637	0,44	ABC1	0,26	45.816,00	0,72	0,09
Macul	5	73	1014	686	0,68	ABC1	0,31	18.475,00	0,69	0,06
Quinta Normal	5	87	843	530	0,63	C3	0,43	24.719,00	0,71	0,08
Estación Central	4	83	845	465	0,55	C3	0,28	11.422,00	0,8	0,16
Ñuñoa	4	50	1828	868	0,47	ABC1	0,04	13.213,00	0,73	0,08
La Florida	2	23	1188	735	0,62	ABC1	0,32	10.357,00	0,7	0,07
Recoleta	2	41	1322	706	0,53	ABC1	0,57	7.767,00	0,76	0,12
Huechuraba	1	38	1091	980	0,9	ABC1	0,01	6.333,00	0,67	0,09
La Cisterna	1	11	864	490	0,57	C3	0,27	5.848,00	0,67	0,07
Providencia	1	39	2849	1139	0,4	ABC1	0,06	3.524,00	0,71	0,1
San Joaquín	1	12	747	764	1,02	ABC1	0,34	4.545,00	0,74	0,12

According to the Pearson correlation study, analysing the number of real estate sales classified as investification and the different values of amendments in each census tract, a significant correlation is observed between the number of international immigrants and the areas with investification processes ongoing. Table 6.6 shows these results for the total number of

Table 6.6 Table of significant Pearson correlations between areas with investification and urban variables.

Variables	*Correlation (Investificated Quantity)*	*Sig.*	*Cov.*	*N*
Investificated Quantity	1,0000	0,0000	0,0034	1644
Immigrants	,531**	0,0000	0,0031	1644
Shop (or Grocery store to be more precise)	,292**	0,0000	0,0019	1644
Hopsital	,260**	0,0000	0,0023	1644
Police	,259**	0,0000	0,0026	1644
Metro Station	,252**	0,0000	0,0022	1644
Museum	,237**	0,0000	0,0025	1644
Fire stations	,231**	0,0000	0,0021	1644
Administrative Services	,225**	0,0000	0,0014	1644
Book shop	,200**	0,0000	0,0007	1644
Plaza	–,200**	0,0000	–0,0015	1644
Supermarket	,200**	0,0000	0,0011	1644
Pharmacy	,189**	0,0000	0,0007	1644
Cinema	,188**	0,0000	0,0020	1644
Library	,181**	0,0000	0,0013	1644
Clinique	,171**	0,0000	0,0015	1644
High-School	,164**	0,0000	0,0013	1644
University	,161**	0,0000	0,0004	1644
Working Age between 15 & 64	,155**	0,0000	0,0012	1644
Medical Centre	,152**	0,0000	0,0005	1644
Veterinary	,152**	0,0000	0,0016	1644
Sport facility	–,145**	0,0000	–0,0012	1644
Bank	,135**	0,0000	0,0003	1644
Bus Stop	,093**	0,0001	0,0007	1644
Elementary School	,068**	0,0060	0,0004	1644
Age between 6 & 14	–,067**	0,0069	–0,0008	1644
Infants	,056*	0,0226	0,0004	1644
Hardware Store	–,050*	0,0419	–0,0005	1644
Street Market	–,050*	0,0419	–0,0005	1644
Sport Court	,049*	0,0475	0,0004	1644
Kindergarten	0,0214	0,3852	0,0002	1644
Elderly over 65 years old	–0,0330	0,1815	–0,0004	1644
General Practice	–0,0401	0,1041	–0,0003	1644
Park	–0,0008	0,9727	0,0000	1644
Religious facility	–0,0065	0,7910	0,0000	1644

areas studied, where immigrants show the highest level of dependence on the number of investified dwellings per census tract. This is followed by shops, hospitals, police stations and metro stations. All these are factors of urban centrality in Santiago. On the other hand, a negative correlation is observed in plazas, sport facilities and infants living in the area. These results seem to be counterintuitive, given that these urban functions or demographic compositions tend to be well valued by investors, but it could be because the presence of these amenities also implies a greater presence of visitors to the neighbourhood and abundant pedestrian movement on the pavements. Certainly, this finding would require further investigation through fieldwork data.

Finally, from an exploratory study of descriptive statistics to compare rental prices for census tracts categorised by levels of investification, it is obtained that in the presence of higher levels of investification, the rental value tends to homogenise, thus generating greater certainty about the expected returns on investment in the sector. This can be seen in Table 6.7, where the average price per square metre of rental housing in areas without investment represents only 24.7% of the value of those areas with high investment and 25.4% of those areas with medium investment. There is also a significant statistical jump between areas with low investification and areas with medium investification. The rental prices in medium and high investificated areas have low variability, which reduces financial uncertainty when investing in properties located in these urban areas.

Some significant findings can be determined from this analysis. First, it confirms the existence of this investification process in the city of Santiago and indicates the associated tendency to push the rental value of properties, even when these values exceed the purchasing power of the residents of each neighbourhood. The findings are indicative that the buy-to-let phenomenon tends to increase the price of housing beyond the neighbours' effective ability to pay, which, on the one hand, puts undue pressure on the household economy and, on the other, leads to the development of neighbourhood gentrification. A second finding is the dependence between the presence of immigrants and the areas of investment. This can have different interpretations that need a specific deepening, but some initial hypotheses of the phenomenon can be associated with labour precariousness that urges immigrants to rent places in well-located areas to reduce mobility costs, preferring communes with good social assistance programmes for migrants. The deregulation of the housing market in Chile in the face of the recent increase in migration considers the shortage of rental housing in central locations a unique opportunity to generate business associated with renting, as has already occurred in other migratory moments in Chile's history.[36] This would be reinforced by the fact that immigrants might be more

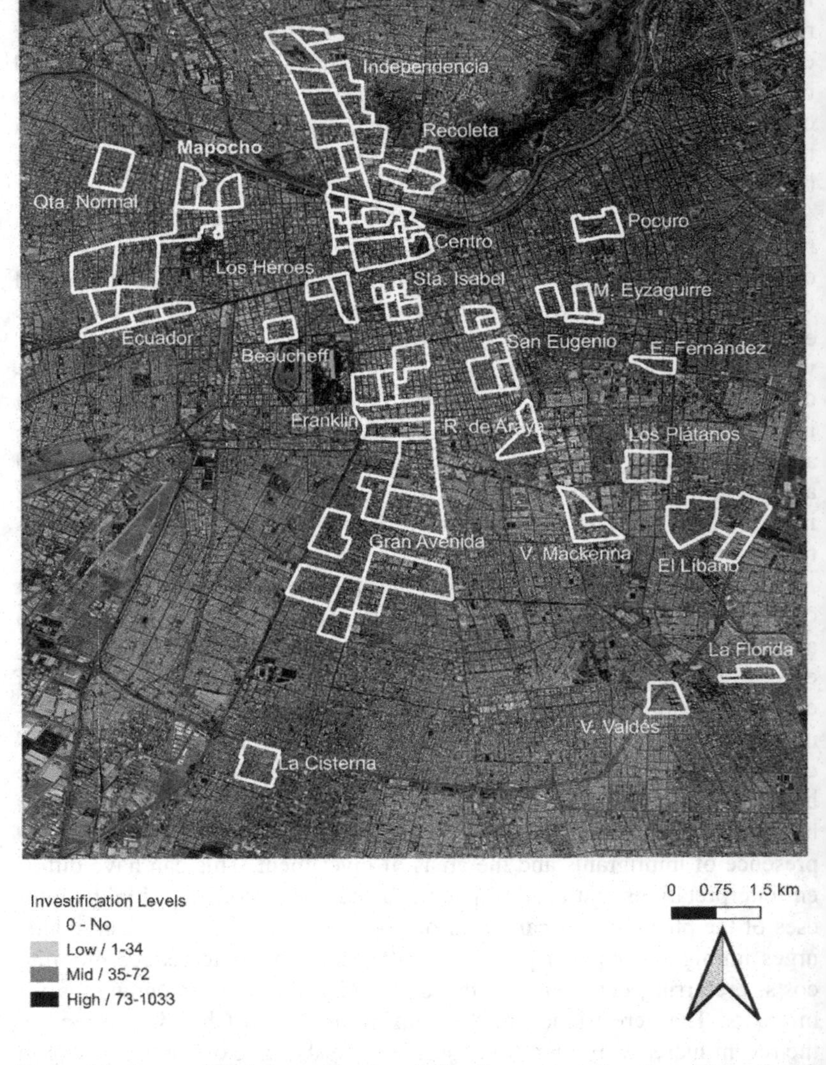

Figure 6.2 Areas of high investification, indicating sector of each neighbourhood observed.

Table 6.7 Descriptive statistics for the exploration of the relationship between rental price per square metre and levels of investification by census tract.

Letting Price per SQM Statistics	*No Investif.*	*Low*	*Mid*	*High*
Mean	0.0629119	0.1362214	0.2481563	0.2546935
Mean Std. Error	0.0047087	0.0051435	0.0082101	0.0060000
95% Confidence Interval for Mean Lower Bound	0.0536659	0.1261288	0.2317973	0.2427356
95% Confidence Interval for Mean Upper Bound	0.0721580	0.1463140	0.2645153	0.2666515
5% Trimmed Mean	0.0484765	0.1236745	0.2487903	0.2568223
Median	–	0.1589680	0.2452083	0.2562911
Variance	0.0144560	0.0281488	0.0050555	0.0026640
Std. Deviation	0.1202332	0.1677761	0.0711017	0.0516138
Minimum	–	–	–	0.0223256
Maximum	0.9380769	2.9464500	0.4579817	0.3566667
Range	0.9380769	2.9464500	0.4579817	0.3343411
Interquartile Range	–	0.2432558	0.0744824	0.0587529
Skewness	1.9879253	4.9554939	– 0.4069237	– 1.1753274
Skewness Std. Error	0.0957097	0.0749884	0.2774001	0.2791970
Kurtosis	5.0556395	74.1941389	3.6310288	4.5486495
Kurtosis Std. Error	0.1911296	0.1498369	0.5482113	0.5516835
Huber's M-Estimation	0	0.1282570	0.2450703	0.2569207

inclined to share housing to reduce costs, given that in the sectors investigated it was found that the situation of housing with *allegados* living at each house was relatively high in relation to Greater Santiago. Finally, it is confirmed that in census tracts with high investification, uncertainty about the rental value of housing is reduced. These areas tend to be located near important road corridors in the city, areas where urban planning regulations allow verticalisation processes with housing buildings and where there is demand for new housing. In order for the investification to take place, there needs to be an increase in the demand for rental housing. In this case, the increase in immigrants can be seen as a source of income, but so is the accelerated increase in the price of housing in relation to household income, where fewer and fewer households can afford to own and must rent. While this allows for the development of cash cows at present, the future scenario is complex, especially after the pandemic, where the casualisation of labour in the middle sectors could generate unrecorded effects on the urban way of life, influenced by teleworking, weak social security and changing views on the space where people will want to confine themselves in the future in the face of the potential "coronaexodus" from cities.[37] This could imply an increase in the concentration of home ownership in high-income sectors, with simultaneous debt capacity, thus undermining the security of housing tenure for middle-income sectors or those without financial debt capacity.

Letting processes are quite a thing in political economy. However, the data of letting prices is not public, and housing prices are not regulated because in Chile, it is unconstitutional to regulate the supply and demand (at least till 2022). The issues related to investification require a closer view from authorities to understand the phenomenon and analyse how this affects the secure tenure of housing. In our view, acknowledging how the neoliberal nature of urban policy and its social effects operate is fundamental as a start to then create a mechanism for tracing its fluctuations. So far, speculative behaviour in the housing market is allowed by the libertine housing policy. In our perspective, this libertinage starts with the neglected approach of the state to land use and its market-oriented way to define land policy.

Notes

1 Rolnik, R. *La guerra de los lugares: la colonización de la tierra y la vivienda en la era de las finanzas* (Santiago: LOM Ediciones, 2017).
2 Salazar, Gabriel. The "Social Blowout" in Chile: A Historical Overview. *Ciper*, 2019. https://ciperchile.cl/2019/10/27/el-reventon-social-en-chile-una-mirada-historica/.
3 Manuel B. Aalbers, The Potential for Financialization, *Dialogues in Human Geography* 5, no. 2 (2015): 214–19, https://doi.org/10.1177/2043820615588158; Manuel B. Aalbers, Financial Geographies of Real Estate and the City Financial Geography Working Paper # 21, 2019, 1–46.
4 Cattaneo Pineda, R. A. Real Estate Investment Funds and Private Housing Production in Santiago de Chile: A New Step towards the Financialisation of the City? *EURE* (Santiago) 37, no. 112 (2011): 5–22.
5 David Harvey in this interview: www.arquine.com/david-harvey/
6 Valencia, Marco. The Free Market City. *Diseño Urbano & Paisaje Magazine* 3, no. 7 (2006).
7 In the constitutional amendment to Law 16.615, the following addition is made: "The law shall establish the way to acquire, use, enjoy and dispose of property and the limitations and obligations that ensure its social function and make it accessible to all. The social function of property includes everything required by the general interests of the State, public utility and health, the best use of productive sources and energies in the service of the community and the improvement of the living conditions of the common inhabitants". This was key to land reform as well.
8 Vergara Perucich, F. and Boano, C. Under Scarcity, is Half a House Enough? Reflections on Alejandro Aravena's Pritzker. *Architecture Review* 21, no. 31 (2016): 37–46. En: doi:10.5354/0719-5427.2016.42516
9 Vergara-Perucich, Francisco. *Urban Design Under Neoliberalism*, 1st ed (New York: Routledge, 2019). En: https://doi.org/10.4324/9780429203268.
10 Donoso, F. and Sabatini, F. Santiago: Real Estate Company Buys Land. *EURE Magazine* 7, no. 9 (1980).

11 Trivelli, P. *Elementos teóricos para el análisis de una nueva política de desarrollo urbano* (Santiago de Chile: Institute of Urban Development Planning, 1981).
12 Bohoslavsky, Juan Pablo et al. *Economic Complicity with the Chilean Dictatorship: an Unequal Country by Force* (Santiago: LOM Ediciones, 2019).
13 Monibot, George. Neoliberalism—The Ideology at the Root of All Our Problems. *The Guardian*, April 2016, pp. 1–5. www.theguardian.com/books/2016/apr/15/neoliberalism-ideology-problem-george-monbiot.
14 Vergara-Perucich, José Francisco and Aguirre, Carlos. Investification in Latin America: Problematising the Letting Market in the Chilean Case, *Habitat and Society*, no. 12 (2019): 1–20.
15 Data from Central Bank, https://si3.bcentral.cl/siete/secure/cuadros/home.aspx
16 www.msn.com/es-cl/noticias/chile/tu-bolsillo-aumenta-morosidad-de-los-chilenos/vp-AABz4mn.
17 Rolnik, Raquel Rolnik, Late Neoliberalism: The Financialization of Homeownership and Housing Rights, *International Journal of Urban and Regional Research* 37, no. 3 (2013): 1058–66, https://doi.org/10.1111/1468-2427.12062.
18 Universidad Alberto Hurtado, *Encuesta Chile Dice* (Hurtado: Universidad Alberto Hurtado, 2017).
19 Nicolás Navarrete and Pablo Navarrete, 'Moving "Away" from Opportunities: Homeownership and Labor Market', Working Paper, 2016.
20 David Harvey, Seventeen Contradictions and the End of Capitalism, *The White Review*, no. 150 (2014): 1–2, https://doi.org/10.1017/CBO9781107415324.004.
21 José Francisco Vergara-Perucich y Carlos Aguirre-Nuñez, Housing Prices in Unregulated Markets: Study on Verticalised Dwellings in Santiago de Chile, *Buildings* 10, no. 1 (2019): 6. https://doi.org/10.3390/buildings10010006.
22 José-Francisco Vergara-Perucich, ¿Qué tan caro es vivir en las capitales regionales? La necesidad de descentralizar las mediciones sobre el costo de vida en Chile, en *El nuevo orden regional. Construcción Social y Gobernanza del Territorio*, ed. Verónica Fuentes, Egon Montecinos, y Pedro Güell, 1st ed (Valdivia: Universidad Austral de Chile, 2020), 145–58.
23 José-Francisco Vergara-Perucich y Carlos Aguirre Nuñez, Inversionistificación en América Latina: problematización del mercado de arriendo para el caso chileno, *Hábitat y Sociedad*, n. 12 (2019): 11–28, https://doi.org/10.12795/HabitatySociedad.2019.i12.02.
24 Kath Hulse and Margaret Reynolds, 'Investification: Financialisation of Housing Markets and Persistence of Suburban Socio-Economic Disadvantage', *Urban Studies* 55, no. 8 (2018): 1655, https://doi.org/10.1177/0042098017734995.
25 Farha, Leilani, *Report of the Special Rapporteur on Adequate Housing as a Component of the Right to an Adequate Standard of Living, and on the Right to Non-Discrimination in This Context, on Her Mission to Chile* (vol. A/HRC/37/5; United Nations: New York, 2018.
26 Cédric Durand, *Fictitious Capital. How Finance Is Appropriating Our Future* (London and New York: Verso Books, 2017).
27 Raquel Rolnik, *La Guerra de Los Lugares. La Colonización de Tierra y La Vivienda En La Era de Las Finanzas* (Santiago: LOM Ediciones, 2017).
28 Caitlin Buckle et al., Marginal Housing during COVID-19, *AHURI Final Report*, no 348 (2020): 1–55, https://doi.org/10.18408/ahuri7325501.

29 José Francisco Vergara-Perucich, Juan Correa-Parra, y Carlos Aguirre-Nuñez, The Spatial Correlation between the Spread of Covid-19 and Vulnerable Urban Areas in Santiago de Chile, *Critical Housing Analysis* 7, no. 2 (2020): 21–35, https://doi.org/10.13060/23362839.2020.7.2.512.
30 Diego Gil McCawley, The Political Economy of Land Use Governance in Santiago, Chile and Its Implications for Class-Based Segregation, *SSRN* 47, no 1 (2012): 119–64, https://doi.org/10.2139/ssrn.2144538.
31 Daniel Santana-Rivas, 'Geografías Regionales y Metropolitanas de La Financiarización Habitacional En Chile (1982–2015): ¿entre El Sueño de La Vivienda y La Pesadilla de La Deuda?', *Eure*, 2020, https://doi.org/10.4067/S0250-71612020000300163.
32 Neil Brenner, Peter Marcuse, and Margit Mayer, *Cities for People, Not Ofr Profit* (New York and Oxon: Routledge, 2012).
33 Loretta Lees, Tom Slater, and Elvin K Wyly, 'Gentrification', 2008, 310, https://doi.org/10.4324/9780203940877.
34 Felipe Yaluff, *Los Secretos de La Inversión Inmobiliaria* (Santiago: Editorial Un Nuevo Día, 2016).
35 Juan Correa-Parra, José-Francisco Vergara-Perucich, y Carlos Aguirre-Núñez, Towards a Walkable City: Principal Component Analysis for Defining Subcentralities in the Santiago Metropolitan Area, *Land* 9, no. 11 (2020): 1–22.
36 Rodrigo Hidalgo-Dattwyler, *La Vivienda Social En Chile y La Construcción Del Espacio Urbano En El Santiago Del Siglo XX* (Santiago: RIL Editores, 2019).
37 Ricardo Greene, Lucía de Abrantes, y Luciana Trimano, Nos/otros: Fantasías geográficas, fricciones y desengaños, *ARQ (Santiago)*, n. 106 (2020): 92–103, https://doi.org/10.4067/S0717-69962020000300092.

7 Land and speculation

The price of land puts pressure on real estate mechanics in the context of liberalised markets, such as the Chilean one. To understand the origin of this discussion, we would like to propose an analogy. Suppose you want to attend a concert in high demand. You would probably not be surprised to find that the price of tickets will be determined by how well you can watch and listen to the show. This would translate into a criterion of location, where the most expensive tickets would be those where you can watch and listen the best and the cheapest, obviously, would be those where you can watch and listen the worst. If it were, moreover, a market determined solely by supply and demand, the best-located tickets would systematically rise in price—since they are more in demand—until the maximum willingness to pay is achieved. However, if we can offer buyers the possibility of paying for their tickets by credit card, the higher prices can be accessed through the acquisition of debt, putting upward pressure on prices and allowing new growth gaps to appear for ticket prices based on the speculation of new concerts with similar demand. As you can imagine, we are not just talking about one concert, but about the city. However, if it is possible to attend several concerts in a year, it is unlikely that we will move house so frequently. This is the underlying problem generated by the real estate mechanics that we will discuss in this chapter: speculating access to urban goods maximises the commercial value of location in the form of exchange value by capturing consumer benefits, but this undermines the attributes associated with the architectural and technical project.

Uses, intensities of uses and what individuals can afford to pay ultimately determine the value of land. This is a triggering discussion for the disciplinary field of urban economics. In 1819, David Ricardo[1] proposed that land does not have a value in itself, but depends on a rent transfer originating in how much the buyer can pay in relation to the value generated by the location of the land in the territory. Subsequently, von Thünen[2] established a basic principle of compensation between transport costs and land

DOI: 10.4324/9781003348771-7

rent, which in summary, states that land closer to territorial centres tends to have higher values because it saves time and energy used to travel to them. Therefore, for a costly location, it is better to densify it in terms of functions and housing in order to obtain higher returns on land investments. Mills and Hamilton[3] showed that in a monocentric city, where all jobs are concentrated in the business and service centre, population and functional density decreases as the distance from the centre increases, following a negative exponential function.

For real estate developers in Chile, there is a direct relationship between the price of land and the projects they design. Feasibility studies for this type of initiative begin by identifying the price of land and its exploitation potential in order to calculate the profitability that a residential project can generate. From the land, the real estate developer adds factors: construction costs, marketing costs, permit processing, sale time and price to be set for the units in order to pay for the investment, plus the respective surplus of monetary profits for investors. Thus, the price of housing is set by the financial capabilities of the demanders. Specifically, in the case of Chile, this price has an upper limit determined by the debt capacity. By this logic, the price of land is the first factor that pushes up the price of housing. In addition, the regulatory conditions establish its design, typology and construction quality.

In simple terms, whoever wants to buy housing buys what they can, but developers set prices according to what someone is willing to pay for housing, regardless of whether the buyer is a person in a housing shortage, an Arab sheikh or a doctor interested in generating income from real estate. As the market is liberalised in the case of Chile, bidders can come from anywhere while housing cannot be anywhere. Therefore, real estate developers exploit the monopolistic condition of the land of the projects they develop through price. Thus, location has an exchange value and with it a political economy of urban space.

Land is a primordial element in the neoliberal sphere, where private property shapes the city. The land and its location also establish values based on the attributes of the surroundings, such as accessibility to services, transport and green areas. These factors influence the price.[4] The owner of the land can apply economic speculation on a potential use associated with competition with other land, depending on its location in the city. In short, the exchange value of land is shaped by this monopolistic condition. This is manifested in the decoupling between the price of land and other productive factors of construction, such as the price of materials.

To illustrate these claims, we conducted an empirical analysis in the city of Santiago. To test the above discussion at the community level, we established a database containing the price of land for each of the census tracts, spatially continuously distributed on the basis of discrete witnesses of real estate transactions. In addition, data on housing prices and construction costs were obtained from the information centre of the Chilean Chamber of Builders.[5] For the analysis, we considered the period from 2008 to 2018. We argue that monopolistic equilibrium should be expressed in the maximisation of location attributes, through an increase in the price of land versus the cost of materials in high-rise construction. Even within this process, given that there is a capture of exchange value by the price of real estate—along with the price of land—the increase in values should respond spatially by replicating the socio-spatial segregation of the city. These impacts on housing and the city can be observed by indexing the price variations of land, high-rise housing and materials[6] for 34 communes over the last ten years (Figure 7.1). At first glance, it is possible to observe how in all cases, the

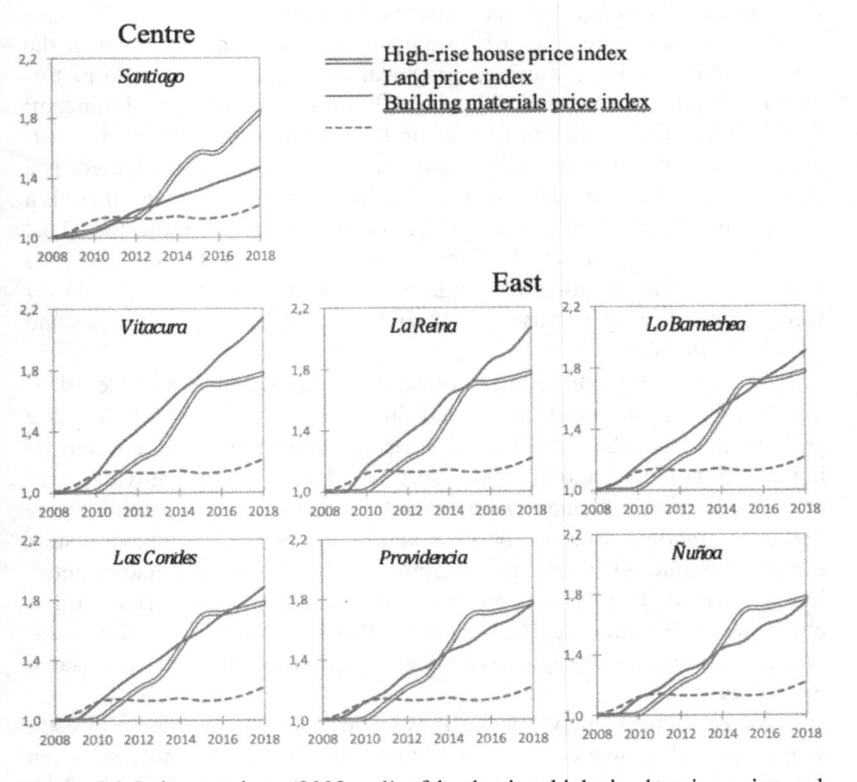

Figure 7.1 Index numbers (2008 = 1) of land price, high-rise housing price and price of construction materials for 34 communes of Greater Santiago.

variation in the price of high-rise housing appears completely decoupled from the variation in the cost of building materials. On the contrary, the relationship between the cost of building materials and the increase in land prices—although it is practically always coupled—does not do so in the same way in all communes, nor in the larger territorial groupings.[7] These results illustrate that the price of land varies according to the monopolistic exploitation of the location, which is part of a speculative process and reflects the socio-residential segregation that it reproduces in its real estate dynamics.

The decoupling between the price of high-rise housing and the price of materials is real. This opens a necessary discussion on the technical quality of housing construction in Chile. It is probably not surprising to note that one of the main arguments put forward by the real estate industry[8] to limit the restrictions associated with construction and energy improvements is associated with the increase in construction costs as a *driver* of the increase in the price of housing. But this impact is almost nil.

The increase in the price of housing is related to the variation in the price of land. Looking at the geographical zoning of the Santiago metropolitan area, there are some differences, but they are not very significant: 84% increase in the price of housing in the central area, 78% for the eastern area, and 69% for the entire western arc. On the other hand, there is a greater disparity in the variation of land prices, especially because there is a resolution at the commune level for the construction of the indicator. Thus, price increases range from 113% in Vitacura, 88% in Las Condes and 75% in Ñuñoa, within the so-called "high income cone" of Santiago, to lower income districts with increases of 34% in Cerro Navia, 23% in Cerrillos and 16% in La Pintana.

This suggests that rather than a direct relationship between house prices and land prices, we must go beyond this pairing to understand this as a problem of class urbanism.[9] The strongest relationship is seen between rising land prices and socio-economic sectors, which is consistent with a process of capturing exchange value through monopolistic competition in the real estate market. This can be observed by comparing the above results at the commune level with the mapping of the Territorial Socio-material Indicator (ISMT)[10] used as a *proxy* for socio-economic level. Here, a first clue appears that validates the hypothesis that land price values and socio-economic status are highly correlated and spatially distributed in the same way (Figure 7.2).

If the relationship between the ISMT and land price variation already suggested the existence of a strong correlation, it is even more so when incorporating the distance to the central business district (Table 7.1). This was located in the new financial district of Santiago,[11] located at the

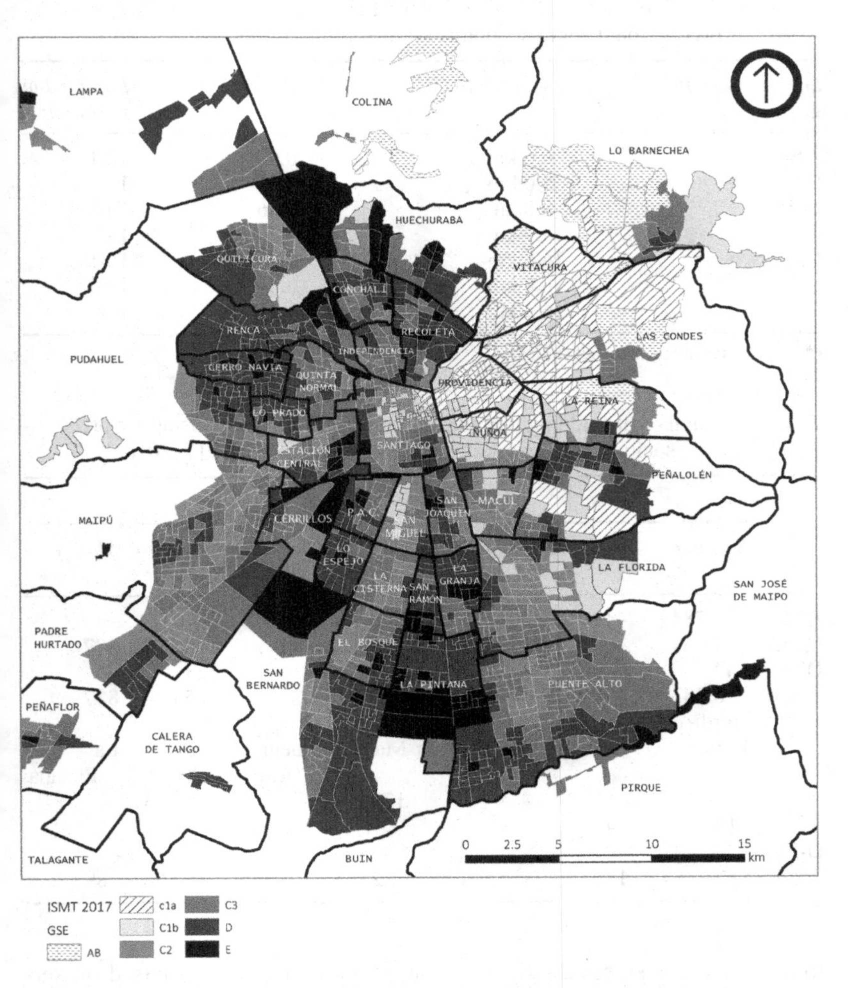

Figure 7.2 Territorial Socio-material Indicator (ISMT) expressed in the predominant socio-economic sector based on 2017 census information for the Santiago metropolitan area.

confluence of the communes of Providencia and Las Condes and popularly known as *Sanhattan*. The values show a relationship between the predominant socio-economic sector, the variation of land prices in the period and the ISMT. A significant segregation of sectors associated with distance from the central business district is observed, with the upper AB and upper-middle classes being better able to compete and position themselves closer

Table 7.1 Average values of indicators for the Santiago metropolitan area, by predominant socio-economic sectors in the census tract.

Socio-economic group	*Distance to CBD*	*ISMT**	*Land value variation*
AB	6.6 km	0.96	1.81
C1a	5.0 km	0.93	1.86
C1b	8.0 km	0.86	1.63
C2	13.9 km	0.76	1.47
C3	16.6 km	0.64	1.51
D	16.8 km	0.54	1.54
E	16.4 km	0.47	1.5

(*) Territorial Socio-material Indicator

Table 7.2 Number of building permits for new multi-family housing by predominant socio-economic sectors in the census tract for 2017–2018.

	Centre	*East*			*North-West*		
	Santiago	*Vitacura*	*Las Condes*	*Ñuñoa*	*Recoleta*	*Estación Central*	*Maipú*
ABC1	3.784	417	694	8.134	–	–	–
C2-C3	4.654	–	80	1.328	683	5.077	887
D-E	344	–	–	–	–	723	–
Total	8.782	417	774	9.462	683	5.800	887
	North-West		South-West				
	Renca	Cerrillos	San Joaquín	Macul	Puente Alto	San Miguel	La Pintana
ABC1	–	–	–	172	–	1.973	–
C2-C3	112	141	–	1.975	1.266	3.525	89
D-E	70	–	–	832	–	–	–
Total	182	141	–	2.979	1.266	5.498	89

to the central business district, consistent with an economy based on services and commerce.[12]

At this point, it is worth asking how real estate activity—in terms of housing production—dialogues in this relationship between land price values and socio-economic level. A study of building permits for new multi-family housing (i.e., more than three floors) shows the concentration of supply in the middle socio-economic levels C2 and C3, with 54% of the total, but increasing to 70% if we discount the "high income cone", corresponding to the communes of the eastern zone (Table 7.2).

A similar conclusion is obtained when comparing building permits by commune, where 73% are concentrated in four communes—from highest to

Table 7.3 Number of building permits for new multi-family housing in different municipalities by year for the period 2008–2018.

	Centre	*East*			*North-West*		
	Santiago	*Vitacura*	*Las Condes*	*Ñuñoa*	*Recoleta*	*Estación Central*	*Maipú*
2008	17.562	499	1.853	2.902	400	2.789	1.060
2009	15.701	397	1.842	809	0	1.216	735
2010	1.973	117	1.219	804	291	192	–
2011	6.234	248	1.996	2.649	147	1.231	–
2012	2.730	481	3.419	2.576	164	2.596	90
2013	7.109	1.837	907	3.493	–	2.161	152
2014	10.844	888	3.235	3.009	231	7.030	302
2015	3.993	597	3.129	6.360	461	11.310	454
2016	7.860	220	1511	2.898	368	10.979	411
2017	5.484	399	1.068	6.181	366	4.305	639
2018	3.298	18	675	3.281	317	1.495	248
Total	82.788	5.701	20.854	34.962	2.745	45.304	4.091

	North-West		*South-West*				
	Renca	*Cerrillos*	*San Joaquín*	*Macul*	*Puente Alto*	*San Miguel*	*La Pintana*
2008	566	150	439	1.860	129	2.667	–
2009	993	–	–	278	74	958	1.299
2010	61	288	286	169	–	972	120
2011	810	–	360	177	–	562	–
2012	488	1.058	–	554	–	1.071	65
2013	146	–	375	218	–	1.424	–
2014	–	263	–	2.507	467	3.942	193
2015	1.088	832	1.299	1.336	238	4.934	393
2016	142	–	517	1.609	98	6.384	–
2017	182	141	–	1.193	1.094	5.251	–
2018	–	–	–	1.786	172	247	89
Total	4.476	2.732	3.276	11.687	2.272	28.412	2.159

lowest: Santiago, Estación Central, Ñuñoa and Las Condes—and dropping to 65% if the last one is discounted. Given that the entire period is shown, it is interesting to observe the moments of contraction and expansion of the real estate industry, highlighting the impacts of the *subprime* mortgage crisis at the beginning of the economic cycle and subsequently a period of growth between 2015 and 2017, particularly for the communes of Estación Central and Ñuñoa. The latter situation seems interesting, since both communes have attracted public attention in recent years due to the media coverage of their densification and building typologies that have populated these areas.

Returning to the original discussion, what is generated here is a competition between monopolists, who change their markets by differentiating themselves from each other from their basic product to the associated services, to the imaginary of urban housing as a symbolic component[13] and to the means of commercialisation. This monopolistic competition configures a real estate offer with levels of differentiation at the level of market approaches, which has contributed to consolidate an imaginary private market for high-rise housing in some communes of Santiago, and for low-rise housing in others.

According to what has been explained throughout this chapter, the different actors in the real estate sector—rentistas—present a behaviour that can be characterised as monopolistic competition, in which equilibrium is produced with the valuation of a complex asset that is capable of establishing a differential valuation of its attributes. However, as has already been explained, the monopolistic power of land has favoured the generation of real estate mechanics in which the locational attributes are strengthened and prioritised over those linked to the architectural and technical project. In our opinion, this whole structure of interactions that we have described is also reflected in the public debate that has been triggered in the light of the social outburst of 18 October 2019. This is because the discussion on housing—in terms of architectural and technical quality—has been largely invisible, above other concerns linked to urban segregation and mobility. While it seems urgent and a priority to us to give back to the city of Santiago its capacity to transport and connect millions of people (between peripheral communes and the central business district, for example), it seems pertinent to us to try to contribute to the understanding of the whole phenomenon. Land is a political economy factor and needs to be incorporated into comprehensive analyses such as those suggested in this book. We believe that starting with an understanding of how real estate speculation processes are produced and reproduced gives us a basis for explaining how certain urban impacts have been able to turn the city of Santiago into a sounding board for public unrest.[14]

Notes

1 Ricardo, David. *On the Principles of Political Economy and Taxation* (London: John Murray, 1819).
2 Thünen, J. H. Von. *Der Isolierte Staat (The Isolated State)* (Hamburg: Perthes, 1826). In: https://doi.org/Cited By (since 1996) 13rExport Date 3 February 2012.
3 Mills, E. and Hamilton, B. W. Urban Economics. *Studies in the Structure of the Urban Economy* (Scott Foresman: Glenview, 1984).
4 José Francisco Vergara-Perucich. Determinantes urbanos del precio de la vivienda en Chile: una exploración estadística, *Urbano* (2021): 40–51.
5 CChC. Indicators. Information Centre, 2019. www.cchc.cl/centro-de-informacion/indicadores/

6 The latter is assumed to be relevant as construction is a material-intensive activity with economies of size (the larger the project, the greater the bargaining power for material costs).
7 The so-called zones in Figure 1, as defined by CChC (2019). *Op. cit.*
8 Encinas, Felipe, Marmolejo-Duarte, Carlos, Wagemann, Elizabeth and Aguirre, Carlos. Energy-Efficient Real Estate or How It Is Perceived by Potential Homebuyers in Four Latin American Countries. *Sustainability* 11, no. 13 (2019): 3531. En: https://doi.org/10.3390/su11133531.
9 Lefebvre, Henri, *La Révolution Urbaine* (Paris: Gallimard, 1970).
10 Developed by the Observatorio de Ciudades UC based on 2017 Census data for the indicators of schooling of the head of household, materiality of the dwelling, housing conditions and overcrowding.
11 Consistent with the analysis of Garreton, Matías. City Profile: Actually Existing Neoliberalism in Greater Santiago. *Cities* 65 (2017): 32–50. En: https://doi.org/10.1016/j.cities.2017.02.005.
12 In this sense, it has been observed that, if we start from a monocentric model—in which all employment is concentrated in the central business district—both economic activity and population density decrease as the distance to the central business district increases, following a negative exponential function.
13 In the terms proposed by Bourdieu, Pierre. *Las estructuras sociales de la economía* (Buenos Aires: Manantial Editorial, 2001).
14 Vergara Perucich, Francisco. Acts of Dissent as Democratising Urbanism: Political Space in Santiago de Chile. *Urbe. Revista Brasileira de Gestão Urbana* 11 (2019).

8 Pandemic and political economy of housing

The COVID-19 pandemic brought the world to a halt and made it possible to observe it at the height of crisis. This is a trial that will judge how prepared nations are for the most important global challenge facing the world—the climate emergency. The results, so far, show that we are not prepared. Even the most developed nations have not managed to confront the pandemic with much level of success, apart from New Zealand and Iceland. The worst is yet to come, as the worldwide infection has spread to many nations in the global south, particularly Latin America, and climate change effects are more severe every year. According to data from Johns Hopkins University, at the end of July 2020, three countries in Latin America were among the top ten countries with the greatest number of accumulated infections worldwide: Brazil (2nd), Peru (7th), and Chile (8th), with 2,552,265; 400,683; and 353,536 cases, respectively (Johns Hopkins University, 2020). To paraphrase Giorgio Agamben,[1] the virus will make us see ourselves and realise how much time we have lost, trapped by goals and objectives that mattered little when it came to deciding between life and death. There is no certainty about which path the world will take after the pandemic. Some, such as Slavoj Zizek, argue that it is time to take a new direction toward post-capitalist scenarios, and others like Byung-Chul Han argue that the bio-political consequences of the pandemic will result in an Orwellian state. Judith Butler emphasises the liberal crisis that the coronavirus has unveiled, which should produce a stronger social vocation for states,[2] while Naomi Klein emphasises that this pandemic is an opportunity for the corporate political elite to implement measures favourable to their interests while the world is distracted by the contagion[3]. These contradictions are being starkly revealed in cities during this pandemic.

DOI: 10.4324/9781003348771-8

This chapter seeks to illustrate how the coronavirus pandemic revealed some urban development contradictions and priorities cities must work on to adopt as a new code of practice during the current health crisis and climate emergency. We analysed empirical evidence that shows how socio-spatial inequality is a health threat illustrated especially by the course of the pandemic in cities of the global south. The aim is to contribute to the development of a revolutionary approach to urban thinking and especially on the political economy of housing as a way to drive changes to foster fairer urban spaces. The case we studied is the Metropolitan Area of Santiago (AMS), given that it presents ideal socio-spatial characteristics for analysis, with very high levels of socio-economic inequality and residential segregation,[4] as illustrated in Figure 8.1. It is also the capital of Chile, a country with one of the highest levels of confirmed cases of COVID-19 per 100,000 inhabitants in the world at the time this test was run.[5]

As most of the chapters of this book, this one took a quantitative approach with variable analysis. The variables were all related to social determinants of health in terms of housing, public transportation, and city infrastructure, and the most relevant were reviewed to determine what other spatial variables might help to explain the increase of cases and the dynamics of infection in various communes of the Santiago Metropolitan Area between 30 March 2020 and 14 August 2020. To carry out this analysis, social determinants of health that corresponded to housing, transportation and urban infrastructure were classified, resulting in a list of 20 variables to be studied. A principal component analysis was applied to these variables. This statistical technique was used to reduce the set of variables and synthesise them into new groups that were not correlated with each other. This method was used in the early stages of statistical analysis to explore and select variables that would remain in the next stage of the study to generate explanatory groups against other variables. In this initial exploration, the variables related to the increase in the cases of COVID-19 per commune were not yet incorporated.

Kaiser-Meyer-Olkin Measure of Sampling Adequacy = 0.552 (significant). Sig = 0.000 (relevant).

5 groups of variables were derived from the analysis, defined by correlations between factorial scores of each of the factors and the initial variables, the results of which are placed with a limit of the value 0.4 (Table 8.1). The groups of variables are described below:

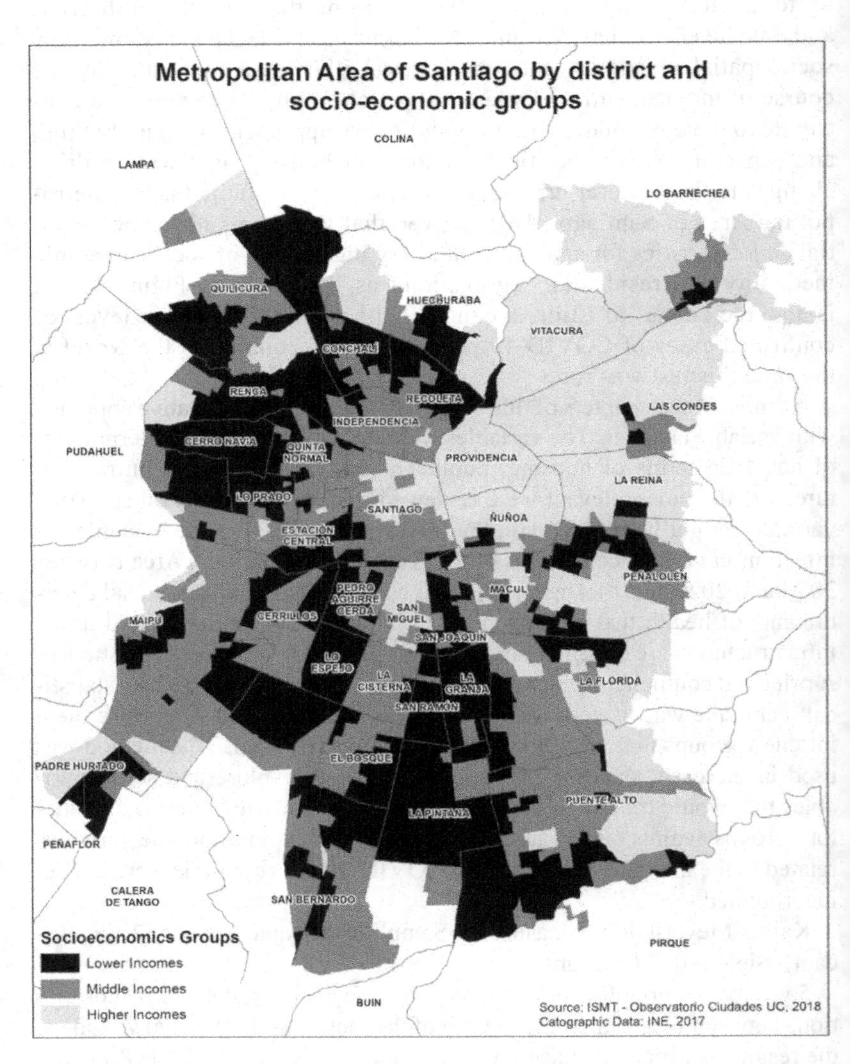

Figure 8.1 Metropolitan Area of Santiago by districts and socio-economic groups.

Source: Authors based on Observatorio de Ciudades UC and Instituto Nacional de Estadísticas.

Table 8.1 Principal component analysis applied to 20 variables related to social determinants of health related to housing.

Rotated Component Matrix[a]

Variables	*Components*				
	FAC1_ Housing based on BID Rent	*FAC2_ Household Vulnerability*	*FAC3_ Amenities*	*FAC4_ Subjective Marginality*	*FAC5_ Municipal Assets*
Less than 2.5 km to sport facilities	.001	−.005	.971	.130	.018
Less than 2.5 km to health facilities	−.013	.061	.950	.120	.012
Less than 2.5 km to educational facilities	−.010	.013	.956	.079	.022
Less than 2.5 km to shops facilities	−.053	−.026	.948	−.056	−.053
Crime in neighbourhood	−.033	.053	.282	.896	−.070
Broadband internet access	.113	−.129	.020	−.850	.151
Workers leaving the commune to their job places	.722	.640	.012	−.106	.003
Bus stop per commune (number)	.255	.872	−.118	−.013	.180
Precarious housing	−.163	.783	.137	.271	−.173
Overcrowding	.074	.955	.092	.053	−.110
Clean energy heating	.292	.161	283	−.747	−.256
Elemental workers (Appendix 1)	−.078	.925	−.038	088	.209
Trips to the CBD of the Metropolis	.875	.388	−.032	−.081	−.015
Destination diversity in the commune	.448	−.340	.207	.079	−.639
Green spaces (parks and plazas) with maintenance	.442	−.132	.060	.237	.694
Letting price to income ratio	−.735	.064	−.031	.112	.093
Average housing prices	.782	−.388	.003	−.101	.094

Rotated Component Matrix[a]					
Variables	*Components*				
	FAC1_ Housing based on BID Rent	*FAC2_ Household Vulnerability*	*FAC3_ Amenities*	*FAC4_ Subjective Marginality*	*FAC5_ Municipal Assets*
Annual Temperature Range	−.632	.291	.297	−.056	−.396
Highly qualified workers	.886	.352	−.032	−.198	−.019
Safe access to tap water	−.031	.054	.077	.078	.829
Mobile data access	.207	−.175	−.015	−.779	−.050
Garbage accumulation	.128	.029	.212	.859	−.036

Factor 1 (Fac1): Housing based on BID rent:	William Alonso's theory of bid rent, in which the influence of distance to the city's main work and commercial area on the cost of living was used to classify housing.
Factor 2 (Fac2): Household vulnerability:	Households were rated according to significant presence of overcrowding, where people work outside their commune of residence and may, for instance, need to travel to the city's central business district (CBD).
Factor 3 (Fac3): Amenities:	Neighbourhoods and communes were rated by how close inhabitants were to sports facilities, educational centres, health centres and shops.
Factor 4 (Fac4): Subjective marginality:	Neighbourhoods that have the perception of exclusion due to the presence of crime, micro-waste dumps, poor access to the internet and networks and the need to use unclean means of heating were identified.
Factor 5 (Fac5): Municipal assets:	Neighbourhoods were classified by management of the communal land, represented by the security of access to drinking water, diversification of goods and services within the commune and number of green areas per inhabitant.

Once the variables to be analysed were defined, a correlation chart was made that identified the concentration of COVID-19 cases by date and by commune with the chosen variables. Pearson's correlation coefficient was used,[6] which is a measure of linear dependence between two variables that do not have the same unit of measurement. This allows an index of the degree of relationship between two compared variables to be developed, and the relationship between two quantitative variables can be graphed, which is what was done in this study. Pearson's correlation coefficient indicates whether or not the social determinants of health in terms of housing, public transportation, and city infrastructure (using the five factors) have influenced COVID-19 expansion in the AMS.

To determine which city and housing explanatory variables were most appropriate to account for the increase in cases of COVID-19, a multiple linear regression was applied.[7] This statistical technique applies a mathematical model of adjustment between a variable under study (cases of COVID-19 by commune and their progress in days) and a set of variables used to explain the behaviour of the variable under study (in this study, the factors resulting from the principal component analysis). The stepwise analysis will make it possible to define whether the statistical model is relevant. It will also help to identify which variables best explain the increase in cases.

The analyses were assisted by the statistical software Microsoft Excel (2016 Redmond, WA: Microsoft Corp.) and IBM SPSS Statistics for Windows, Version 22.0 (Armonk, NY: IBM Corp.). The cartographic evaluations and socio-spatial analysis were developed using ArcGIS (Version 10.5) software (Redlands, CA: Esri, Inc.). The data produced by the Centro Producción del Espacio of the University of the Americas and the Observatorio de Ciudades of the Faculty of Architecture, Design and Urban Studies of the Pontificia Universidad Católica de Chile were used. These urban research centres have official databases with a strong emphasis on geo-referenced documentation from the 2017 Census and complementary data provided by the National Institute of Statistics, Internal Revenue Service, Ministry of Transport and Telecommunications, Ministry of Health, Ministry of Housing and Urban Development, and the Santiago Real Estate Conservative (CBRS) through the data company Inciti.

Among important social determinants of health is the linkage of public health strategies with housing policies that provide security of tenure,[8] which greatly reduces mental illness by allowing people to be confident in calling a specific place home.[9] For Tord Kjellstrom and Susan Mercado, "Urban development and town planning are key to creating supportive social and physical environments for health and health equity".[10] The processes and dynamics of urbanisation have become epidemiological factors that have the ability to promote or stop the spread of pathogens in cities.[11] The recent pandemic has revealed the need to rethink cities so they can incorporate urgent innovations because of climate change. Housing must undergo profound transformations to combat the public health emergency while reducing greenhouse gases.[12] The coronavirus pandemic has revived old questions about problems that have never been resolved. Urbanism without humanitarian or health-care considerations is not new. Almost 100 years ago, Le Corbusier launched a book about the need for transformation of architecture to meet the new requirements of 20th century society, and his disciplinary revolution transformed cities forever.[13] It was after the Spanish flu pandemic that urban architecture inevitably arrived at the port of a new disciplinary revolution, and it is urgent that we reproduce that innovation with interdisciplinary approaches before economic interests take over, following the example of Le Corbusier after the early 20th century pandemic.[14] In this respect, it is worth reviving Carolyn Stephens' question[15]: "Do I study the health of the urban poor and ignore the structural processes that create those health problems and that poverty?" With regard to this approach, the case of Chile is presented as a critical scenario to assess the scope of potential consequences that would result from further postponing this discussion.

Chile's neoliberal urbanism has demonstrated its inability to generate just cities.[16] Current policies have produced unequal spaces with high levels of residential segregation,[17] in which the quality of urban public goods differs drastically according to neighbourhood characteristics.[18] In addition, the difficulties faced by households in affording high housing prices[19] have resulted in a significant increase in informal settlements in the country's main cities.[20] In an urban development organised for income rather than social objectives,[21] market rules end up making it difficult to improve the spatial quality of households in an equitable manner, and this also produces inequalities in the production of habitat.[22] The coronavirus pandemic has shown that these social injustices have spatial correlation and are themselves health threats.[23] Chile has an urban configuration marked by inequalities that can facilitate the spread of infectious diseases.[24] In the 1920s, Chilean architecture and city infrastructure underwent a substantial change, driven in part by the conditions of urban unhealthy behaviour and social injustice that existed at the time.[25] One hundred years later, the factors that require urban re-evaluation from the social and public health points of view are once again combined with an environmental factor and demand interdisciplinary approaches to an urgent urban revolution.[26]

Specifically in the framework of the current pandemic, progress is needed between epidemiology and urban studies. To this end, there are some research experiences known as socio-spatial epidemiology. This type of study focuses on investigating why certain population groups experience dissimilar health outcomes despite sharing biological aspects. This is due to social differences, such as residential segregation.[27] This disciplinary articulation uses quantitative methods of analysis from spatial statistics to geo-reference results and understands health problems from a territorial perspective. The methods most used in this novel disciplinary field refer to the study of correlations and multiple linear regressions. Continuing with the proposal indicated in the previous section of methods, the first exercise consisted of broadening a Pearson's correlation graph according to the increase of cases by communes and the social determinants of health, which were grouped into five factors. The result can be seen in Figure 8.2.

Figure 8.2 clearly indicates that the development of novel coronavirus cases moved toward households with greater vulnerability, as measured by the working conditions of their inhabitants, while the case load in the best-located households in relation to urban centrality (Figure 8.3) under William Alonso's bid rent theory was decreasing.[28] To a lesser extent, the social determinants of health outside housing also display a similar trend, in which municipalities with good capital reduced their infections while communes

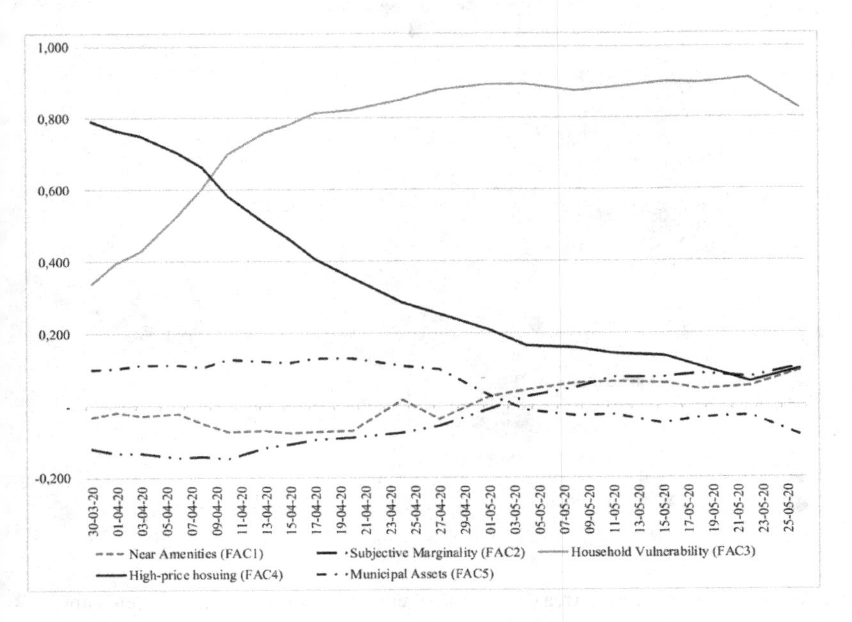

Figure 8.2 Pearson's correlation coefficients compare the components of social determinants of health in housing and urban development with the daily increase of cases per commune.

Figure 8.3 Metropolitan Area of Santiago weighted distance to employment centres.
Source: Authors based on Observatorio de Ciudades UC and Instituto Nacional de Estadísticas.

with greater subjective marginality and public (not private) infrastructure experienced increases in cases.

In Santiago, the expansion of COVID-19 cases originated in a sector of the city with good urban quality, and then concentrated on the sectors where urban conditions were more precarious. Although the correlation does not imply causality, the relationship between these variables is high as measured by Pearson's coefficient. To confirm the validity of this model, a multiple linear regression was developed, which yielded the following results.

The stepwise linear multiple regression model (Table 8.2) demonstrates that Factor 1, Factor 3, Factor 4 and Factor 5 do not provide the statistical significance to explain the case increases in communes of the Santiago Metropolitan Area. Factor 2, which describes aspects of housing that expose vulnerability, is the independent variable that best explains the increase in COVID-19 cases. This is also relevant because of the high statistical significance of the model (adjusted r2 of 0.874). The specific elements that make up Factor 2 are presence of elementary workers in the household, amount of overcrowding, amount of precarious housing, number of bus stations per person, and number of workers travelling to other communes for work.

The Ministry of Health of Chile itself has recognised the impact of the "social determinants" on the spread of COVID-19 to the most vulnerable sectors of the Santiago Metropolitan Area, identifying them as "the quality of housing, lack of work and poverty". It is clear, then, that in the movement of COVID-19 from the more affluent sectors of Santiago to the periphery, the SARS-CoV-2 virus, which is responsible for COVID-19 infections, found particularly suitable physical and material conditions to rapidly increase the number of infections. Among these, the characteristics of the houses themselves are particularly interesting, since public policies intended to improve the quality of the thermal envelope in urban housing have been systematically postponed. An example of this has been the process of updating the current thermal regulations. The current regulations have been defined as weak and insufficient by international organisations, such as the Organisation for Economic Co-operation and Development (OECD) and the World Bank, and this criticism has been endorsed by numerous national studies. These updates have been on hold for 13 years and must be done in order to meet the country's commitments to decarbonise the built environment by 2050. In terms of precariousness or low quality according to its characterisation in the 2017 Chile Population and Housing Census, the role that housing has played in this pandemic and in the behaviour of other endemic respiratory viruses in the country makes the implementation of all public building policies urgent. When poor housing is combined with overcrowding, a "perfect storm" is created for spreading the SARS-CoV-2

Table 8.2 Results of multiple linear regression model.

Model Summary[b]

Model	R	R Square	Adjusted R Square	Std. Error of the Estimate	Change Statistics					Durbin-Watson
					R Square Change	F Change	df1	df2	Sig. F Change	
1	.937[a]	.878	.874	11.60213	.878	230.337	1	32	.000	2.135

[a] Predictors: (Constant), Fac2 (Household vulnerability)

[b] Dependent Variable: Progress of the infection in the city until 14 August

ANOVA[a]

Model		Sum of Squares	df	Mean Square	F	Sig.
1	Regression	31005.535	1	31005.535	230.337	.000b
	Residual	4307.500	32	134.609		
	Total	35313.035	33			

[a] Dependent Variable: Progress of infection in the city until 14 August

[b] Predictors: (Constant), Fac2 (Household vulnerability)

Coefficients[a]

Model		Unstandardised Coefficients		Standardised Coefficients	t	Sig.	95,0% Confidence Interval for B		Correlations			Collinearity Statistics	
		B	Std. Error	Beta			Lower Bound	Upper Bound	Zero-order	Partial	Part	Tolerance	VIF
1	(Constant)	52.903	1.990		26.588	.000	48.850	56.956					
	FAC2_ Household Vulnerability	30.652	2.020	.937	15.177	.000	26.538	34.766	.937	.937	.937	1.000	1.000

Source: [a] Dependent Variable: Progress of infection in the city until 14 August

Source: Authors elaboration.

virus in houses, among family circles and in the immediate environment (neighbourhood) that surrounds them.

International literature has also revealed the role that urban mobility has played in the spread of the virus,[29] which depends not only on the modes of transport, but also on the form and organisation of the city itself. Although, at times, the Metropolitan Area of Santiago has shown signs of development patterns that tend toward polycentrism, it still has a powerful monocentric component around the already-described CBD.[30] The fact that workers commute from the communities where they live to the CBD as a result of the concentration of work opportunities is a relevant finding that explains the increase in COVID-19 infections. The fact that the direction of the spread also correlates positively with some bus stops suggests that the vast majority of these trips are by public transportation. The latter can also be understood as a vector of contagion since its use has presented many difficulties when it comes to social distancing.

These results open up reflections on how the inequalities of social determinants of health related to housing and urban planning, even in a metropolis in one of the most advanced nations of the global south, affect the resolution of urgent health risks. Decades of neoliberal policies and a false process of modernisation are reviving situations that were thought to have been overcome. At a time of fundamental crisis, the city is once again plagued by "common pots" (ollas comunes), a form of social organisation among communities that ensures people can eat, and grassroots organisations trying to help, but they have no support from this absent state and a nation without a social welfare system. Marginalisation has not been eradicated from the nation, and neoliberalism has only been effective in postponing social integration to the interior of the country,[31] so that those who have the resources to take refuge and protect themselves in their neighbourhoods and good quality housing will prosper during a crisis situation, as is dictated by the law of the strongest. Meanwhile, the majority of the damage is in more vulnerable sectors, such as the communes of Puente Alto, Maipú, La Florida, Peñalolén and La Pintana, and in the overcrowded spaces in the commune of Santiago[32] (Figure 8.4).

Although the virus was introduced to the city via the highest income sector, its sustained reproduction in vulnerable sectors shows the validity of a precariópolis as an effective viral reproducer. The high mobility and dependence on travel for work or formalities of people living in the most vulnerable and peripheral sectors to the city centre and CBD also add to the labour informality of this social group, which prevents proper confinement. We understand the precariópolis as monofunctional and segregated areas of the city where there is greater socio-economic homogeneity among the inhabitants, and where they live in small houses

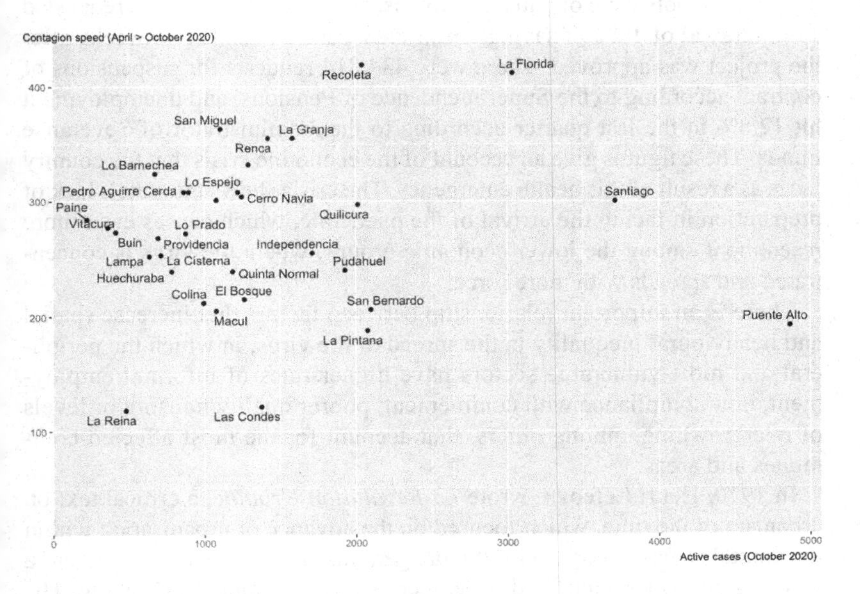

Figure 8.4 Scatter plot of communes according to predictability values of the applied multiple linear regression model.

poorly constructed.[33] Although the precariópolis was initially thought to be confined to the edges of the city, it has been penetrating inward because of a housing production model that allows for the existence of 19 m^2 typologies sold at very high prices as a result of their location.[34] The precariópolis, once an exclusive product of the state, is now also a product of the property market. Greater Santiago is beginning to convert enormous neighbourhoods into exclusively residential spaces and tends to concentrate its central functions in the vicinity of the upper quarter, which is known as the centre-east sector of the city. The price of housing, educational, health and leisure services make it impossible for lower-income households to live in these areas, and this is where there was least damage due to the new coronavirus. Santiago is the epitome of an unequal city. It is so segregated that one sector is like the Elysium in Blomkamp or Zalem in Kishiro, where a limited number of people can afford to live, while the rest of the city offers the opposite of a quiet and prosperous urban life. The pandemic has allowed urban studies experts to contemplate this urban crisis, and this has opened up opportunities to rethink the city in terms of spatial justice.

Common pots were organised in towns, and 80% of pensioners requested a withdrawal of 10% of savings from their pension funds two weeks after the project was approved. There were 434,814 requests for suspensions of contract according to the Superintendence of Pensions, and unemployment hit 12.8% in the last quarter according to the Administrator of Severance Funds. These figures give an account of the economic crisis that the country faces as a result of this health emergency. The crisis shows the state's lack of preparation in facing the arrival of the pandemic, which causes even more resentment among the lower economic groups, where the virus is concentrated and spreads with more force.

There is an important relationship between factors that increase spatial and behavioural inequality in the spread of the virus, in which the peripheral and most vulnerable sectors have higher rates of informal employment, non-compliance with confinement, poorer quality housing or levels of overcrowding, among others, that account for the most affected communes and areas.

In 1970, Henri Lefebvre wrote *La Revolution Urbaine*, a critical text on urbanism of the time, which focused on the advance of urbanisation within the capitalist framework, with the dangers that this implied, to generate a more egalitarian society and a democratically distributed urban life. His main concern was quantifiable: when the world reached 100% urbanisation, it would be very difficult to reverse the effects of class urbanism that had shaped the capital city.[35] In Chile, urbanisation has already reached 90%, so following Lefebvre's thesis, there is very little room for change. An urban revolution, where the disciplines of urbanism manage to divest themselves of the ballasts of capital, is the best and fastest way toward a society where everyone has access to the benefits of urban life. Before the beginning of the pandemic, Chile had embarked on a path of structural political change following a nationwide social crisis, known as the social explosion of 18 October 2019. One of the results of that revolt was the beginning of a constitutional process. In the face of a new constitution, rethinking the city as a collectively produced entity that exceeds the responsibility for urban planning alone seems to be key. It is one of the lessons learned from this study: a segregated, unequal city, guided by the rules of capital, is a health risk and needs interdisciplinary approaches. As such, ceasing to see the city as a space for extracting value and beginning to situate it as a human right is urgent in the face of the even greater challenges that climate change will bring.

It is not the first time that an infectious disease has made us rethink the way we design our cities. The French hygiene movement at the end of the 19th century was one of the main driving forces behind the guiding principles of the modern architecture movement, and it was a search for bright and clean spaces that inspired the first part of the career of the influential Le Corbusier. It is entirely possible that we face a disciplinary revolution in the making that will transform the way our cities are designed. As the city is a social reflection, a change in its organisation for health purposes will necessarily imply a rethinking of the social relations that give life to the built environment. With this, it is not surprising that a new social contract is imperative as part of creating a more prosperous future for civilisation. In Chile, this new social contract could be embodied in the new constitution, but surely, new international agreements will bring that agenda to life. From the data provided by our research, it is irresponsible to continue postponing the city's reorganisation to ensure fair access to a good life for the majority of the population. The pandemic has been a preliminary warning of the great anthropogenic threat that is climate change.

On this last point, urban planning cannot neglect its healthcare role and, in this respect, the importance of moving toward overcoming profit-oriented architecture and breaking with the harmful logic of the financialisation of urban processes cannot be overlooked. From the economic paradigm of urban development, it is necessary to move once and for all to an integrative paradigm, where the priorities are placed on energy poverty, climate change, and the spatial elimination of social inequality. These are major challenges, but to continue postponing their solutions seems foolish.

The empirical analysis shows that this is the general framework necessary for the transformation of Santiago and cities and society in general. It reinforces what was happening before the pandemic. Social inequalities are everywhere in the built space. It is also present in the way vulnerable inhabitants of the city, who are not necessarily quarantined, relate to each other in their search for daily sustenance.

The Chilean model is an example of the neoliberal reforms in Latin America and has structural flaws in the formation of the most dispossessed neighbourhoods, which constitute the housing complexes promoted by the state. In these sectors, the premium is on hidden informality, which are precarious settlements and places for renting rooms or bedrooms—very typical of Chilean reality.

Notes

1 Giorgio Agamben, 'Monólogo Del Virus', *Lundimatin*, 27 March 2020, https://lundi.am/Monologo-del-Virus-2853.

2 Zizek, Han and Butler in: Pablo (editor) Amadeo, *Sopa de Wuhan. Pensamiento Contemporáneo En Tiempo de Pandemias*, ed. Pablo Amadeo (Santiago: ASPO (Aislamiento Social Preventivo y Obligatorio), n.d.).

3 Marie Solis, 'Coronavirus Is the Perfect Disaster for "Disaster Capitalism"', *Vice*, 13 March 2020. Available at: https://www.transcend.org/tms/2020/03/coronavirus-is-the-perfect-disaster-for-disaster-capitalism/ (Accessed 22-11-2022)

4 Alejandra Rasse, Segregación residencial socioeconómica y desigualdad en las ciudades chilenas, *Serie Documentos de Trabajo PNUD* 4 (2016); Camila Cociña, Habitar desigualdades: Politicas urbanas y el despliegue de la vida en Bajos de Mena, *Serie Documentos de Trabajo PNUD- Desigualdad*, n. 2016/05 (2016): 1–21.
5 Johns Hopkins University, 'World Trends', COVID-19 Dashboard by the Center for Systems Science and Engineering (CSSE) at Johns Hopkins University (JHU), 2020, https://coronavirus.jhu.edu/map.html.
6 A. Berni et al., Effect of Vascular Risk Factors on Increase in Carotid and Femoral Intima-Media Thickness. Identification of a Risk Scale, *Atherosclerosis* 216, no. 1 (2011): 109–14.
7 Gregory D. Andranovich and Gerry Riposa, *Doing Urban Research* (Thousand Oaks: SAGE Publications, 1993).
8 Lauren A. Taylor et al., Leveraging the Social Determinants of Health: What Works?, *PLoS ONE* 11, no. 8 (2016): 1–20.
9 Cheryl Forchuk, Kevin Dickins, y Deborah J. Corring, Social Determinants of Health: Housing and Income, *Healthcare Quarterly (Toronto, Ont.)* 18 (2016): 27–31, https://doi.org/10.12927/hcq.2016.24479.
10 Tord Kjellstrom y Susan Mercado, Towards Action on Social Determinants for Health Equity in Urban Settings, *Environment and Urbanization* 20, no. 2 (2008): 538, https://doi.org/10.1177/0956247808096128.
11 Emilie Alirol et al., Urbanisation and Infectious Diseases in a Globalised World, *The Lancet Infectious Diseases* 11, no. 2 (2011): 131–41, https://doi.org/10.1016/S1473-3099(10)70223-1.
12 Alice Sverdlik, Ill-health and Poverty: A Literature Review on Health in Informal Settlements, *Environment and Urbanization* 23, no. 1 (2011): 123–55, https://doi.org/10.1177/0956247811398604.
13 Le Corbusier, *Towards a New Architecture*, *Towards a New Architecture* (New York: Dover Publicat, 1986), https://doi.org/10.1016/b978-0-85139-652-1.50006-8.
14 J N Hays, *Epidemics and Pandemics. Their Impacts on Human History* (Santa Barbara, CA, Denver, Oxford: ABC Clio, 2005).
15 Carolyn Stephens, Revisiting Urban Health and Social Inequalities: The Devil is in the Detail and the Solution is in All of Us, *Environment and Urbanization* 23, no. 1 (2011): 37, https://doi.org/10.1177/0956247811398588.
16 Francisco Vergara-Perucich, *Urban Design Under Neoliberalism: Theorising from Santiago* (Oxon and New York: Routledge, 2019); Camillo Boano y Francisco Vergara-Perucich, *Neoliberalism and Urban Development in Latin America*, 2017; Rodrigo Hidalgo Dattwyler, Voltaire Christian Alvarado Peterson, y Daniel Santana Rivas, La espacialidad neoliberal de la producción de vivienda social en las áreas metropolitanas de Valparaíso y Santiago (1990–2014): ¿hacia la construcción idelógica de un rostro humano?, *Cadernos Metrópole* 19, no. 39 (2017): 513–35, https://doi.org/10.1590/2236-9996.2017-3907; Cristhian Figueroa Martínez et al., Creating Inequality in Accessibility: The Relationships between Public Transport and Social Housing Policy in Deprived Areas of Santiago de Chile, *Journal of Transport Geography* (2017): 0–1, https://doi.org/10.1016/j.jtrangeo.2017.09.006; Encinas et al., Speculation, Land Rent, and the Neoliberal City. Or Why Free Market is Not Enough.
17 Alejandra Rasse, 'Segregación Residencial Socioeconómica y Desigualdad En Las Ciudades Chilenas', *Serie Documentos de Trabajo PNUD*, 2016/04; Camila

Cociña, 'Habitar Desigualdades: Politicas Urbanas y El Despliegue de La Vida En Bajos de Mena', *Serie Documentos de Trabajo PNUD* (2016): 1–21. Rasse, 'Segregación Residencial Socioeconómica y Desigualdad En Las Ciudades Chilenas'.

18 Yasna Cortés and Victor Iturra, 'Market versus Public Provision of Local Goods: An Analysis of Amenity Capitalization within the Metropolitan Region of Santiago de Chile', *Cities* 2019, https://doi.org/10.1016/j.cities.2019.01.015.

19 Jose Francisco Vergara-Perucich y Camillo Boano, Vida urbana neoliberal: estudio de factores de jerarquización y fragmentación contra el derecho a la ciudad en Chile, *Revista de Direito da Cidade* 11, no. 3 (2019): 426–52, https://doi.org/10.12957/rdc.2019.38541.

20 Pablo Flores, Campamentos, la cara visible de la exclusión, *Radio Universidad de Chile*, 2018, http://radio.uchile.cl/2017/10/23/aumento-de-campamentos-la-cara-visible-de-la-exclusion-habitacional/; Francisco Vergara-perucich y Camillo Boano, El precio por el derecho a la ciudad ante el auge de campamentos en Chile The rise of urban slums in Chile, *Revista AUS* 26 (2018): 51–57, https://doi.org/10.4206/aus.2019.n26-09.

21 José-Francisco Vergara-Perucich et al., *Contribución a La Economía Política de La Vivienda En Chile* (Santiago: RIL Editores, 2020).

22 Felipe Encinas y Carlos Aguirre, Sustentabilidad y mercado: aproximaciones desde la promoción inmobiliaria, *ACE: Architecture, City and Environment*, 2017, https://doi.org/10.5821/ace.12.35.5141; Felipe Encinas et al., Energy-Efficient Real Estate or How It Is Perceived by Potential Homebuyers in Four Latin American Countries, *Sustainability* 11, no. 13 (2019): 3531, https://doi.org/10.3390/su11133531.

23 Felipe Irarrazaval y Felipe Irarrazaval, La no tan loca geografía del COVID 19 en Chile, *Journal of Latin American Geography* (2020): 1–5.

24 José-Francisco Vergara-Perucich, Juan Correa, and Carlos Aguirre-Núñez, *Atlas de Indicadores Espaciales de Vulnerabilidad Ante El COVID-19 En Chile* (Santiago: Centro Producción del Espacio, 2020).

25 Rodrigo Hidalgo-Dattwyler, *La Vivienda Social En Chile y La Construcción Del Espacio Urbano En El Santiago Del Siglo XX* (Santiago: RIL Editores, 2019).

26 The concept of the Urban Revolution is used in Henri Lefebvre's sense, as a spatial transformation but also in the conceptualisations, epistemologies and disciplinary articulations of the urban area as an object of study that transcends the merely spatial to become interwoven with the social.

27 Andrew B. Lawson et al., *Handbook of Spatial Epidemiology* (Boca Raton: CRC Press, Taylor and Francis Group, 2016).

28 The bid rent theory is a spatial-economic theory that explains how the price and demand for real estate products change as the distance from the central business district (CBD) increases. In Santiago, the CBD is near Line 1 of the Metro system, between Los Heroes and Manquehue.

29 Jason Corburn et al., 'Slum Health: Arresting COVID-19 and Improving Well-Being in Urban Informal Settlements', *Journal of Urban Health* (2020), https://doi.org/10.1007/s11524-020-00438-6; Julio D. Dávila, 'Covid-19, Urban Mobility and Social Equity', *DPU Blog*, 2020, https://blogs.ucl.ac.uk/dpublog/2020/05/04/covid-19-urban-mobility-and-social-equity/.

30 Ricardo Truffello y Rodrigo Hidalgo, Policentrismo en el Área Metropolitana de Santiago de Chile: reestructuración comercial, movilidad y tipificación de

subcentros, *EURE (Santiago)* 41, no. 122 (2015): 49–73, https://doi.org/10.4067/S0250-71612015000100003.

31 Roger Vekemans, Jorge Giusti, and Ismael Silva, *Marginalidad, Promoción Popular e Integración Latinoamericana*, ed. DESAL, Ediciones (Buenos Aires, 1970).

32 For the location of each district and the average income, please check the map in Appendix 1.

33 Rodrigo Hidalgo-Dattwyler, ¿Se acabó el suelo en la gran ciudad?: Las nuevas periferias metropolitanas de la vivienda social en Santiago de Chile, *EURE (Santiago)* 33, no. 98 (2007): 57–75. https://doi.org/10.4067/S0250-71612007000100004.

34 Vergara-Perucich et al. *Contribución a la economía política de la vivienda en Chile* (Santiago: Ril Editores, 2020).

35 Henri Lefebvre, *La Révolution Urbaine* (Paris: Gallimard, 1970).

9 Afterword

Housing as political economy

Political economy studies the relations of production and marketing of goods and services in the light of the legislative, institutional and governmental frameworks that generate livelihoods for the population, with a spatial focus on the distribution and modes of wealth creation.[1] While political economy is often associated with the investigation of macroeconomic factors, a nation's housing and its institutional system or market itself falls into the category of being an integral factor aggregated to the overall economy as a whole. This is especially clear when housing as a system becomes a social problem when entering periods of crisis, as is the situation in much of the world today. If the housing allocation system has been failing for years, the problem is not in individual housing or in the particularities of each household, but in the very design and functioning of the political and economic system that manages the allocation of housing. In the case of Chile, that system is based primarily on the housing market. One of the political economy challenges posed by the housing crisis is the massiveness of the demand for solutions in the face of a system of solution provision that is clearly incapable.[2] This becomes more complex in a model based on tenure as a social aspiration. In the case of Chile, it would be difficult to install a system such as the Austrian one based on leases without evictions, given that the culture in Chile and in Latin America in general aims for housing solutions to become family property rather than state property. It is institutional fragility, along with the recent history of dispossession that prevents, in part, a cultural change in the prioritisation of tenure over other means such as cooperative, public or collective tenure, among others. Solutions can emerge from political economy, specifically through its research methods on social relations. In this book, we have engaged a mixture of political economy methods such as historical analysis, comparative studies, spatial relations and the analysis of social movements and grassroots organisations. In studying the political economy of housing, we are investigating how the state intervenes in the processes of social reproduction,

DOI: 10.4324/9781003348771-9

capital accumulation and the circulation of goods and services, for which the management, organisation and division of land and modes of housing production are critical factors to understand.[3] In its essence, political economy is a framework for the critical investigation of capital.

Housing is a case of political economy, not only because of the captive demand that generates the housing deficit of demographic fluctuations, but also because it is also the object of the creativity of capital. Thus, spatial variations in housing are accompanied by financial variations in investment mechanisms and actors who find ways to extract income from an elementary need of the population, such as a habitat. As Enrique Dussel argues, this creativity of capital is pushing humanity towards the abyss, initially through the exploitation of labour, to which the degradation of the environment was added, and which also includes the reduction of housing space in order to maximise the profitability of investment in urban land, among other possible relationships.[4] At present, housing is a safe haven for financial capital, which, given the abundance of income, sees real estate investment as a space to store its surplus.[5] Faced with the threat of financialisation, the price of housing will not stop rising because there is so much surplus liquidity in the financial system that investors will always find buyers, even if these buyers are not inhabitants but, for example, international investment funds.[6] Finance capital will buy up each other's housing in order to keep it growing, regardless of whether the structural housing deficit continues to increase. The price of housing, in our opinion, is the factor to be observed from a political economy point of view, in relation to the purchasing power of the population. If the mismatch exists, as we have indicated in this book with empirical evidence, political economy must generate tools and inputs to make decisions about it. Avoiding the crisis or helping to resolve it are few of the contributions we seek to make in publishing this book.

This book has presented a set of reflections informed by data produced with academic rigour. We believe that these data allow us to broaden the discussion and ensure that the problem characterised as the political economy of housing overcomes certain constraints imposed by agents with vested interests in the real estate market. While it is true that in any part of the world, the business sectors exert pressure on the productive and legislative processes aimed at regulating their activities, in the real estate market in Chile, this relationship is disturbing. For example, in the National Urban Development Policy, a public policy missive aimed at regulating land, real estate business and urban management, one of the signatories was the Chilean Chamber of Construction, an actor that operates as a manager of business interests. There is no need for malice in this fact, but rather there is a naturalisation of empowering business actors so much so that it is the regulator (the State) itself that invites the regulated (market) to define its regulatory guidelines and sign it, proving that it agrees with the regulation it will be subject to. Certainly, we believe that the actors who participate in the market can give their opinion and influence from the value produced by the practice of the profession, but this is very different from this opinion becoming the norm and the weight of these views being horizontally equated with scientific opinion. The latter is worrying, but also deserves self-criticism.

It is at least questionable that in urban development research processes, with an emphasis on housing, actors from academia are constantly linked to private enterprise, quickly confusing science with consultancy. Clearly, universities and research centres must have relationships and dialogues with actors in society, such as business, but it is also very important to develop certain levels of autonomy that prevent ethical conflicts in the production of knowledge. Otherwise, decision-makers find themselves trapped in a body of knowledge that is often dissonant with each other. Unfortunately, science in Chile today has no *lobbying* power and this is a great disadvantage for the creation of public policies. Under neoliberal logic, these difficulties are highly profitable for big business, which ends up having an impact on the cost of living of citizens. In this ecosystem, the state and the authorities tend to profess a faith in the market, a religious attitude towards a deity that is not invisible or immaterial, but is composed of actors, interests and capital and is recorded in transactions. From our position as social scientists with expertise in the political economy of housing, we invite the reader to look at these phenomena not from this ideological conception, but from the data. It is important to trust methods and results, but it is also important to be critical and rigorous. That is why so many authors have participated in this book, always seeking to maintain an ad hoc rigour to the importance of the subject we are dealing with, with diverse perspectives from different converging experiences.

After decades of a system in which the authorities' decisions were strongly influenced by market actors, today households are confronted with a major problem: the cost of living. Just based on the data presented here, it is clear that housing in Chile has become a luxury commodity, difficult to access for most people and, in the absence of a public housing system, much of the access is conditioned by the purchasing power of households. This purchasing power, in turn, is determined by the ease of access to debt and therefore to financial power. As in the game of *Monopoly*, the bank always wins. Social relations have been transformed, they have taken on a financial form and everything is tradable. In Chile, particularly, we live in a kind of constant trade under the rule of neoliberalism. It has not yet been written what fate has in store for us from the events of 18 October 2019, and three years after that event, the problems that triggered that outburst have not been solved: retirees still have miserable pensions, the cost of living has risen more than ever, wages have not been readjusted and housing continues to rise at faster rates than the average cost of living. Perhaps one of the few advantages of the pandemic was that it forced us to stop life for a while, to look at ourselves as a society and see that inequality really does kill, and very quickly. The new coronavirus imposed a sense of urgency on an urban revolution that cannot be postponed, where the political economy of housing must provide evidence to guide the paths of such disciplinary revolutions.

The neoliberal model has a speculative nature, where the search for the best investment and the highest profit makes agents with significant purchasing power look for safe spaces where capital will not stop growing. In this, urban spaces are tremendously useful for capital, which ends up reproducing socio-spatial segregation, which is reflected in the value of land, housing prices and the way in which the real estate market pushes these assets to become spaces for unregulated commercialisation. This is not aimed at the common good but at the ends of neoliberalism, characterised by a constant growth in the value of capital and the control of social processes based on the profitability of the decisions that are taken, which in this case, refers to housing. In this, it differs from *Monopoly*, because, in the game, everyone starts the game on equal terms. Nothing could be further from the truth, since for the neoliberal model, one of the mechanisms for the generation of wealth lies precisely in the existence of asymmetries.[7]

In this same context, the evidence shatters the myth of the law of supply and demand in the case of housing, where the price is supposedly defined at a precise point of equilibrium between what the entrepreneur seeks to earn and what people can afford to pay. In Chile, most people cannot afford housing without falling into heavy long-term debt, which breaks the supposed market equilibrium. The real estate market is imperfect, so it is not regulated in relation to the optimal point of supply and demand, but is fixed from the supply side (companies), pushing prices up at the expense of demand (consumers), which is disproportionately inflated when in the equation, financial actors disguise themselves as consumers to rent from the use value of housing. In simple words, the real estate market does not care about the entire population's ability to pay in search for universal access to a good without substitutes, but rather sets its prices based on specific portions of the population or institutions that can pay those prices, even if that implies a phenomenon of accumulation of housing by dispossession, gentrification or investification,[8] thus displacing the possibility of everyone having access to housing at a fair price. If we had a real estate market with prices defined at the point of equilibrium in Chile, considering that people do not pay more than 25% of their monthly income for housing ($560,000 according to the

2018 Supplementary Income Survey), with the average housing production cost in Chile of 41 UF/m^2 (including land and utilities of 10%), a person should be able to access a house of 38 m^2 by allocating about $140,000 per month.[9] But these prices are fictitious for the neoliberal ideological framework prevailing in Chile and surely in much of the world. This reasoning does not apply due to the idealisation of the market, which makes it urgent that it be subject to regulations in which households' ability to pay is taken into account in sale or rental prices.

The housing crisis arises partly because the individual outweighs the collective. This is due to interpreting society as if the so-called *homo-economicus* was a real person making decisions with an economic handbook as guide. It is difficult to advance in effective solutions if housing is understood almost exclusively as private property, preventing the development of the city as a whole. After functioning many years following this logic, the housing solutions have proved to not be successful in results nor considering the current complex social scenario in urban areas around the world. Under this individualistic conception, the public is residual—and not the result of collective and integrated planning—where housing is installed without any consideration for its relationship with the city. The private logic of the housing market in Chile relies on a state responsible for providing infrastructure, transport and services, without receiving retribution in return. In contrast, in many other parts of the world, there is what is known as capital gains recovery, that is, what the state does for the public—which increases the value of private assets—is partially recovered to reinvest those resources in new public projects. For example, if I have a house and they build a metro station in front of it, I, as an individual, did nothing to have the station located there; I was just lucky. And I was lucky because my property increased in price by 50% overnight. The capital gains recovery mechanism captures part of that increase for the state in case I sell that property. This helps to prevent strategic infrastructures such as the Santiago Metro from promoting speculative processes to becoming real contributions to social integration. Today, the main role of the Metro is to move people from the peripheries to the city centre, triggering real estate development wherever a station is located. A metropolis as large as Santiago should aim for polycentrality. Given the known scarcity of resources in some peripheral communes, not carrying out capital gains recovery processes is tremendously inefficient and ends up accentuating the difficulties in access to housing as one gets closer to centrality.

Since there is a process of financialisation of housing, it is good to ask what awaits our properties in the future if this phenomenon does not stop. Do design, energy efficiency or architectural attributes matter when the shares are traded on the stock exchange? They probably don't, as long as the property sells and is profitable. Strictly speaking, as long as the property is well located, the architectural and technical design will take a back seat. In fact, there are architectural typologies that are repeated everywhere in Chile, true templates that give little thought to how they are located, what contributions they make to public space and how they adapt to the specific conditions of the territories. This is why investors do not waste much time in designing better housing, but rather seek to maximise profits by building at high speed and selling as quickly as possible at the best value someone can afford. This is called mass production of housing units ready to extract profitability from the investment.

Housing as a human right is an imperative that needs to be advanced more decisively. In the Chilean context, changes occurred with the events of October 2019. We do not know if housing has been incorporated into the debate for a new Chile yet. We do not have certain answers yet, however, we do believe that the possibility of a new Political Constitution of the Republic with housing conceived as a right will be part of this new Chile, in order to begin to move towards spatial justice.

In the current neoliberal context, citizens are confronted with the difficulties of everyday life: old age, illness, job insecurity, indebtedness and access to decent housing. In the process of redefining the institutional framework of the nation, it should be considered a basic duty of the state to solve these problems, accompanying the solutions with robust public policies. We are optimistic about the possibilities for the future, but critical of the steps to be taken. If a new political order is to be implemented with human rights as a priority, housing will have to emerge from its current stagnation to become more humanised. However, the road to this goal is not yet mapped out and we hope that this book will fuel useful discussions to this end.

Notes

1 Vincent W Bladen, *An Introduction to Political Economy*, 2016. www.degruyter.com/doi/book/10.3138/9781442632103.
2 Michael Ball, *Housing Policy and Economic Power: The Political Economy of Owner Occupation* UP 828 (London and New York: Methuen, 1983).
3 Alexander R Cuthbert, *Understanding Cities: Method in Urban Design* (London: Routledge, 2011).
4 Enrique Dussel, '16 Tesis de Economía Política', 2014, 18.
5 Gregory W. Fuller, *The Political Economy of Housing Financialization*, Comparative Political Economy (Newcastle upon Tyne: Agenda Publishing, 2019).
6 Manuel B. Aalbers, *Subprime Cities: The Political Economy of Mortgage Markets, Subprime Cities: The Political Economy of Mortgage Markets*, 2012, https://doi.org/10.1002/9781444347456; Manuel B. Aalbers and Anne Haila, 'The Financialization of Housing: A Political Economy Approach', *Urban Studies* (2018), https://doi.org/10.1177/0042098018759251.

7 Ha-joon Chang, *23 Things They Don't Tell You about Capitalism* (London: Penguin, 2011), 23; Joseph E. Stiglitz, *The Price of Inequality: How Today's Divided Society Endangers Our Future*, 1st ed (New York: W.W. Norton & Co, 2012); Paul Ekins y Manfred A. Max-Neef, eds., *Real-life Economics: Understanding Wealth Creation* (London and New York: Routledge, 1992).
8 Vergara-Perucich, José Francisco and Aguirre, Carlos. Inversionistification in Latin America: Problematising the Rental Market for the Chilean Case. *Hábitat y Sociedad*, no. 12 (2019): 1–20.
9 Hypothetical calculation based on 22 UF/m² in execution (40,128 UF), 5 UF/m² for operating costs associated with the project (9,120 UF), 10 UF/m² on the land transferred to the cost of each dwelling on an 800 m² plot of land (18,240 UF), in a building of eight floors and 48 flats with a sales cost for each flat of 1,558 UF and profits for the seller of close to 7,300 UF, more than two hundred million pesos.

References

A reflection on this extra-disciplinary understanding is developed by journalist Hector Cossio in *El Mostrador* (http://shorturl.at/vCUW1) and the Hyatt Foundation itself in the statement of the arguments for awarding Aravena the Pritzker Prize: So-called incremental housing allows social housing to be built on land that is more expensive but closer to economic opportunities and gives residents a sense of personal investment. At: www.pritzkerprize.com/announcement-ale-jan-dro-ara-ve-na.

A. Berni et al., Effect of Vascular Risk Factors on Increase in Carotid and Femoral Intima-Media Thickness. Identification of a Risk Scale, *Atherosclerosis* 216, no. 1 (2011): 109–14.

Aalbers, Manuel B. and Anne Haila, 'The Financialization of Housing: A Political Economy Approach', *Urban Studies* (2018), https://doi.org/10.1177/0042098018759251.

Alejandra Rasse, Segregación residencial socioeconómica y desigualdad en las ciudades chilenas, *Serie Documentos de Trabajo PNUD* 4 (2016);

Amoako, Clifford Amoako y Emmanuel Frimpong Boamah, Build as You Earn and Learn: Informal Urbanism and Incremental Housing Financing in Kumasi, Ghana, *Journal of Housing and the Built Environment* 32, no. 3 (2017): 429–48, https://doi.org/10.1007/s10901-016-9519-0.

Atria, Fernando, *El otro modelo* (Santiago: Random House Monda- dori, 2013).

Ball, Michael, *Housing Policy and Economic Power: The Political Economy of Owner Occupation* UP 828 (London and New York: Methuen, 1983).

Bladen, Vincent W., *An Introduction to Political Economy*, 2016. www.degruyter.com/doi/book/10.3138/9781442632103.

Boano, Camillo y Francisco Vergara-Perucich, *Neoliberalism and Urban Development in Latin America (London and New York: Routledge, 2017)*.

Bohoslavsky, Juan Pablo, Fernandez, Karinna, Smart, Sebastian (ed). *Economic Complicity with the Chilean Dictator- ship: an Unequal Country by Force* (Santiago: LOM Ediciones, 2019).

Bourdieu, Pierre. *Las estructuras sociales de la economía (Buenos Aires: Manantial Editorial, 2001)*.

Bourdieu, Pierre. *The Social Structures of the Economy*, (Cambridge: Polity Press, 2005).

Brenner, Neil, Peter Marcuse, and Margit Mayer (ed), *Cities for People, Not for Profit (New York and Oxon: Routledge, 2012).*

Burgos, Soledad Burgos et al., Residential typologies in Chilean irregular settlements with precarious housing conditions, *Revista Panameri- cana de Salud Publica* 29, no 1 (2011): 32–40, https://doi.org/dx.doi.org/ S1020-49892011000100005.

Caitlin Buckle et al., Marginal Housing during COVID-19, *AHURI Final Report*, no 348 (2020): 1–55, https://doi.org/10.18408/ahuri7325501.

Camila Cociña, Habitar desigualdades: Politicas urbanas y el despliegue de la vida en Bajos de Mena, Serie Documentos de Trabajo PNUD- Desigualdad, n. 2016/05 (2016): 1–21.

Cattaneo Pineda, Rodrigo. Real Estate Investment Funds and Private Housing Production in Santiago de Chile: A New Step Towards the Financialisation of the City? *EURE* (Santiago) 37, no. 112 (2011): 5–22.

CChC. Indicators. Information Centre, 2019. www.cchc.cl/centro-de-informacion/indicadores/

Cédric Durand, *Fictitious Capital. How Finance Is Appropriating Our Future. (London and New York: Verso Books, 2017).*

Chang,Ha-joon, *23 Things They Don't Tell You about Capitalism* (London: Pen- guin, 2011), 23.

Chovar, Alejandra, and H. Salgado. "¿Cuánto influyen las tarjetas de crédito y los créditos hipotecarios en el sobre endeudamiento de los hogares en chile." Banco Central de Chile, Vol. 8 (2010). Available at: https://www.bcentral.cl/documents/33528/133585/bcch_archivo_140212_es.pdf/16a59094-1e07-9f2e-aa2b-63c4aca06f4b?t=1573288203160 (accessed 22-11-2022)

Comité Hábitat y Vivienda, *Dignidad Humana en el Territorio y la Ciudad (Santiago: Lom Ediciones, 2016).*

Contreras,Dante & Ffrench-Davis, Ricardo. «Policy Regimes, Inequality, Poverty and Growth: The Chil- ean Experience, 1973–2010», Unu-Wider, 2012, www.wider.unu.edu/stc/repec/ pdfs/wp2012/WP2012-004.pdf.

Corburn, Jason, 'Slum Health: Arresting COVID-19 and Improving Well-Being in Urban Informal Settlements', *Journal of Urban Health* (2020), https://doi.org/10.1007/s11524-020-00438-6; Julio D. Dávila, 'Covid-19, Urban Mobility and Social Equity', *DPU Blog*, 2020, https://blogs.ucl.ac.uk/ dpublog/2020/05/04/covid-19-urban-mobility-and-social-equity/.

Correa-Parra, Juan, José-Francisco Vergara-Perucich, y Carlos Aguirre-Núñez, Towards a Walkable City: Principal Component Analysis for Defining Sub-centralities in the Santiago Metropolitan Area, *Land* 9, no. 11 (2020): 1–22.

Cortés, Yasna , and Victor Iturra, 'Market versus Public Provision of Local Goods: An Analysis of Amenity Capitalization within the Metropolitan Region of Santiago de Chile', *Cities* 2019, https://doi.org/10.1016/j.cities.2019.01.015.

Cuthbert, Alexander R, *Understanding Cities: Method in Urban Design* (London: Routledge, 2011).

Cuthbert, Alexander R. *Understanding Cities: Method in Urban Design*. London: Routledge, 2011.

DAHER, Antonio. Territorios de la financiarización urbana y de las crisis inmobiliarias. Rev. geogr. Norte Gd. [online]. 2013, n.56, pp.7–30. http://dx.doi.org/10.4067/S0718-34022013000300002.

Deamer, Peggy Deamer, *Architecture and capitalism: 1845 to the present*, (London: Routledge, 2013).

Donoso, F. and Sabatini, F. Santiago: Real Estate Company Buys Land. *EURE Magazine* 7, no. 9 (1980).

Donoso, Francisco Donoso y Francisco Sabatini, Santiago: empresa inmobili- aria compra terrenos, *EURE: Revista Latinoamericana de Estudios urbanos y Territoriales* 7, no. 20 (1980): 25–51.

Durán, G. and Kremerman, M. Low wages in Chile, 2019. www.fundacionsol. cl/wp-content/uploads/2019/04/Salarios-al-Li%CC%81mite-2017-NV2-1.pdf

Dussel, Enrique, '16 Tesis de Economía Política', 2014, 18.

Edwards, Sebastián. The Reality of Inequality and Its Perception: Chile's Paradox Explained. *Pro-Market*, 19 November 2019. https://promarket.org/ the-reality-of-inequality-and-its-perception-chiles-paradox-explained/

Emilie Alirol et al., Urbanisation and Infectious Diseases in a Globalised World, *The Lancet Infectious Diseases* 11, no. 2 (2011): 131–41, https://doi. org/10.1016/S1473-3099(10)70223-1.

Encinas, Felipe & Carlos Aguirre, Sustentabilidad y mercado: aproximaciones desde la promoción inmobiliaria, *ACE: Architecture, City and Environment*, 2017, https://doi.org/10.5821/ace.12.35.5141; Felipe Encinas et al., Energy- Efficient Real Estate or How It Is Perceived by Potential Homebuyers in Four Latin American Countries, *Sustainability* 11, no. 13 (2019): 3531, https://doi. org/10.3390/su11133531.

Encinas, Felipe et al., Speculation, Land Rent, and the Neoliberal City. Or Why Free Market is Not Enough, *Revista ARQ* 1, no. 102 (2019): 2–15.

Encinas, Felipe, Marmolejo-Duarte, Carlos, Wagemann, Elizabeth and Aguirre, Carlos. Energy-Efficient Real Estate or How It Is Perceived by Potential Homebuyers in Four Latin American Countries. *Sustainability* 11, no. 13 (2019): 3531. En: https://doi.org/10.3390/su11133531.

Farha, Leilani. 2017. *Report of the Special Rapporteur on Adequate Housing as a Component of the Right to an Adequate Standard of Living, and on the Right to Nondiscrimination in this Context, on her Mission to Chile*. New York: United Nations.

Felipe Yaluff, *Los Secretos de La Inversión Inmobiliaria* (Santiago: Editorial Un Nuevo Día, 2016).

Figueroa, Cristhian Martínez et al., Creating Inequality in Accessibility: The Relationships between Public Transport and Social Housing Policy in Deprived Areas of Santiago de Chile, *Journal of Transport Geography* (2017): 0–1, https://doi. org/10.1016/j.jtrangeo.2017.09.006

Flores, Pablo, Campamentos, la cara visible de la exclusión, Radio Universidad de Chile, 2018, http://radio.uchile.cl/2017/10/23/aumento-de-campamentos-la-cara-visible-de-la-exclusion-habitacional/; Francisco Vergara-perucich y Camillo Boano, El precio por el derecho a la ciudad ante el auge de campamen- tos en Chile The rise of urban slums in Chile, Revista AUS 26 (2018): 51–57, https://doi.org/10.4206/aus.2019.n26-09.

Forchuk, Cheryl, Kevin Dickins, y Deborah J. Corring, Social Determinants of Health: Housing and Income, *Healthcare Quarterly (Toronto, Ont.)* 18 (2016): 27–31, https://doi.org/10.12927/hcq.2016.24479.

Fuller, Gregory W. *The Political Economy of Housing Financialization*, Com- parative Political Economy (Newcastle upon Tyne: Agenda Publishing, 2019).

Gasic, Ivo Gasic, Inversiones e intermediaciones financieras en el mercado del suelo urbano. Principales hallazgos a partir del estudio de transacciones deterrenos en Santiago de Chile, 2010–2015, Eure 44, no 133 (2018): 29–50, https:// doi.org/10.4067/s0250-71612018000300029.

Gil McCawley, Diego, The Political Economy of Land Use Governance in Santiago, Chile and Its Implications for Class-Based Segregation, *SSRN* 47, no 1 (2012): 119–64, https://doi.org/10.2139/ssrn.2144538.

Giorgio Agamben, 'Monólogo Del Virus', *Lundimatin*, 27 March 2020, https://lundi.am/Monologo-del-Virus-2853.

Greene, Ricardo, Lucía de Abrantes, y Luciana Trimano, Nos/otros: Fantasías geográficas, fricciones y desengaños, *ARQ (Santiago)*, n. 106 (2020): 92–103, https://doi.org/10.4067/S0717-69962020000300092

Gregory D. Andranovich and Gerry Riposa, *Doing Urban Research* (Thousand Oaks: SAGE Publications, 1993).

Guiloff Titiun, M. Regulatory Expropriation: An Impertinent Doctrine to Control the Imposition of Limits to the Right to Private Property in the Chilean Constitution. *Ius et Praxis* 2, no. 2 (2018): 621–648.

Harberger, Arnold Harberger, Notas sobre los problemas de vivienda y planificación de la ciudad, *Revista AUCA* 37, no. 1 (1979): 39–41.

Harvey, David. Seventeen Contradictions and the End of Capitalism, (Oxford: *Oxford University Press*, 2014).

Harvey, David. *The Enigma of Capital and the Crises of Capitalism* (Madrid: Ediciones Akal S. A, 2012).

Hays, J. *Epidemics and Pandemics. Their Impacts on Human History* (Santa Barbara, CA, Denver, Oxford: ABC Clio, 2005).

Henri Lefebvre, *La Revolución Urbana* (Alianza Editorial, 1972), Madrid.

Hidalgo Dattwyler, R. A.; Paulsen Bilbao, A. G. and Santana Rivas, L. D. Subsidiary Neoliberalism and the Search for Justice and Equality in Access to Social Housing: The Case of Santiago de Chile (1970–2015). *Andamios* 13, no. 32 (2016): 57–81.

Hidalgo Dattwyler, Rodrigo ,Voltaire Christian Alvarado Peter- son, y Daniel Santana Rivas, La espacialidad neoliberal de la producción de vivienda social en las áreas metropolitanas de Valparaíso y Santiago (1990–2014):¿hacia la construcción idelógica de un rostro humano?, Cadernos Metrópole 19, no. 39 (2017): 513–35, https://doi.org/10.1590/2236-9996.2017-3907

Hidalgo Dattwyler, Rodrigo Hidalgo Dattwyler, Alvarado Peterson Christian Voltaire, y Daniel Santana Rivas, La espacialidad neoliberal de la producción de vivienda social en las áreas metropolitanas de Valparaíso y Santiago (1990–2014): ¿hacia la construcción idelógica de un rostro humano?, *Cadernos Metrópole* 19, no. 39 (2017): 513–35.

Hidalgo-Dattwyler, Rodrigo , ¿Se acabó el suelo en la gran ciudad?: Las nue- vas periferias metropolitanas de la vivienda social en Santiago de Chile, *EURE (Santiago)* 33, no. 98 (2007): 57–75. https://doi.org/10.4067/ S0250-71612007000100004.

Hidalgo-Dattwyler, Rodrigo, *La Vivienda Social En Chile y La Construcción Del Espacio Urbano En El Santiago Del Siglo XX* (Santiago: RIL Editores, 2019).

Hidalgo-Dattwyler,Rodrigo, *La Vivienda Social En Chile y La Construcción Del Espacio Urbano En El Santiago Del Siglo XX (Santiago: RIL Editores, 2019).*

Hulchanski, J. David. The Concept of Housing Affordability: Six Contemporary Uses of the Expenditure to Income Ratio, *Housing Studies* 10, no. 4 (1995), https://doi.org/10.1080/02673039508720833

Hulse, Kath & Margaret Reynolds, Investification: Financialisation of housing markets and persistence of suburban socio-economic disadvantage, *Urban Studies* 55, n. 8 (2018): 1655–71, https://doi.org/10.1177/0042098017734995.

Ingevec, *Memoria Anual 2018*, SVS (Santiago: Ingevec, 2018), 68.

Irarrazaval, Felipe, La no tan loca geografía del COVID 19 en Chile, Journal of Latin American Geography (2020): 1–5.

Johns Hopkins University, 'World Trends', COVID-19 Dashboard by the Center for Systems Science and Engineering (CSSE) at Johns Hopkins University (JHU), 2020, https://coronavirus.jhu.edu/map.html.

Kakwani, Nanak Kakwani y Robert J. Hill, Economic Theory of Spatial Cost of Living Indices with Application to Thailand, *Journal of Public Economics* 86, no. 1 (2002): 71–97, https://doi.org/10.1016/S0047-2727(00)00174-2.

Kath Hulse and Margaret Reynolds, 'Investification: Financialisation of Housing Markets and Persistence of Suburban Socio-Economic Disad- vantage', *Urban Studies* 55, no. 8 (2018): 1655, https://doi.org/10.1177/ 0042098017734995.

Kjellstrom, Tord, y Susan Mercado, Towards Action on Social Determinants for Health Equity in Urban Settings, *Environment and Urbanization* 20, no. 2 (2008): 538, https://doi.org/10.1177/0956247808096128.

Lauren A. Taylor et al., Leveraging the Social Determinants of Health: What Works?, PLoS ONE 11, no. 8 (2016): 1–20.

Lawner, Miguel Lawner, *¿Qué hacer?* (Santiago: Carta personal, 2019).

Lawson, Andrew B., *Handbook of Spatial Epidemiology* (Boca Raton:CRC Press, Taylor and Francis Group, 2016).

Le Corbusier, *Towards a New Architecture, Towards a New Architecture* (New York: Dover Publicat, 1986), https://doi.org/10.1016/b978-0-85139-652- 1.50006-8.

Lefebvre, Henri, *La producción del espacio* (Madrid: Captian Swing, 2013)

Lefebvre, Henri, *La Révolution Urbaine* (Paris: Gallimard, 1970).

Leilani Farha, *Report of the Special Rapporteur on Adequate Housing as a Component of the Right to an Adequate Standard of Living, and on the Right to Non-Discrimination in This Context, on Her Mission to Chile* (vol. A/HRC/37/5; New York, 2018). United Nations.

Llanto, Gilberto M. Shelter Finance Strategies for the Poor: Philippines, Environment and Urbanization 19, no. 2 (2007): 409–23, https://doi.org/10.1177/0956247807082821.

Loretta Lees, Tom Slater, and Elvin K Wyly, 'Gentrification', (London: Routledge, 2008).

Manuel B. Aalbers, The Potential for Financialization, Dialogues in *Human Geography* 5, no. 2 (2015): 214–19, https://doi.org/10.1177/2043820615588158; Manuel B. Aalbers, Financial Geographies of Real Estate and the City Financial Geography Working Paper # 21, 2019, 1–46.

Mills, E. and Hamilton, B. W. Urban Economics. *Studies in the Structure of the Urban Economy* (Scott Foresman: Glenview, 1984).

Nicolás Navarrete and Pablo Navarrete, 'Moving "Away" from Opportunities: Homeownership and Labor Market', Working Paper, 2016.

OECD, *OECD Employment Outlook 2018*, 2018, https://doi.org/10.1787/ empl_outlook-2018-en.

OECD, Urban Policy Reviews, Chile 2013, (London: OECD Publishing, 2013)

Pablo Trivelli, Reflexiones en torno a la política Nacional de Desarrollo Urbano, EURE: Revista Latinoamericana de Estudios urbanos y Territoriales 8, no. 22 (1981): 43–64; Antonio Daher, Neo- liberalismo urbano en Chile, Estudios Públicos, 1990, 281–99.

Paredes, Dusan Paredes y Patricio Aroca, Metodología para Estimar un Indice Regional de Costo de Vivienda en Chile *, *Cuadernos de eConomía* 45 (2008): 129–43, https://doi.org/10.4067/S0717-68212008000100005.

Paul Ekins y Manfred A. Max-Neef, eds., Real-life Economics: Understanding Wealth Creation (London and New York: Routledge, 1992).

Paulsen, Alex Paulsen, Negocios inmobiliarios, cambio socioespacial y contestación ciudadana en Santiago Poniente. El caso del barrio Yungay: 200–2013, *La ciudad Neoliberal, gentrificación y exclusión en Santiago de Chile, Buenos Aires, Ciudad de México y Madrid (Santiago: Instituto de Geografía*, 2014).

Peter D. Linneman y Isaac F. Megbolugbe, Housing Affordability: Myth or Reality?, *Urban Studies* 29 (1992): 369–92, https://doi. org/10.1080/00420989220080491.

Pulso. Construcción advierte riesgo de crisis social por creciente costo de las viviendas. La Tercera Pulso, 10 May 2019.

Redding, Stephen Redding, Spatial Income Inequality, *Swedish Economic Policy Review* 12, no. 1 (2005): 29–55.

Rolnik, R. *La guerra de los lugares: la colonización de la tierra y la vivienda en la era de las finanzas* (Santiago: LOM Ediciones, 2017).

Rolnik, Raquel Rolnik, Late Neoliberalism: The Financialization of Home-ownership and Housing Rights, *International Journal of Urban and Regional Research* 37, no. 3 (2013): 1058–66, https://doi.org/10.1111/1468-2427.12062.

Salazar, Gabriel Salazar, *La enervante levedad historica de la clase politica civil (Chile 1900–1973)* (Santiago: Debate, 2015); Gabriel Salazar, *Historia de la acu- mulación capitalista en Chile (apuntes de clases)* (Santiago: LOM Ediciones, 2003).

Salazar, Gabriel, *El poder nuestro de cada día. Pobladores. Historia. Acción popular constituyente* (Santiago: Lom Ediciones, 2016).

Salazar, Gabriel. The "Social Blowout" in Chile: A Historical Overview. *Ciper*, 2019. https://ciperchile.cl/2019/10/27/el-reventon-social-en-chile-una-mirada-historica/.

Samara, Tony Roshan, Anita Sinha, y Marnie Brady, Putting the "Public" Back in Affordable Housing: Place and Politics in the Era of Poverty Deconcentration, Cities 35 (2013): 319–26, https://doi.org/10.1016/j.cities.2012.10.015.

Santana-Rivas, Daniel, 'Geografías Regionales y Metropolitanas de La Financiarización Habitacional En Chile (1982–2015): ¿entre El Sueño de La Vivienda y La Pesadilla de La Deuda?', Eure, 2020, https://doi.org/10.4067/ S0250-71612020000300163.

Socovesa, Memoria Anual 2018, *Memoria* (Santiago: Socovesa, 2018), 57.

Solimano, Andres, Chile and the Neoliberal Trap: The Post-Pinochet Era (Cambridge: Cambridge University Press, 2012)

Stephen G. Cecchetti, Nelson C. Mark, y Robert Sonora, Price Index Convergence among United States Cities, *International Economic Review* 43, no. 4 (2002): 1081–99.

Stephens, Carolyn, Revisiting Urban Health and Social Inequalities: The Devil is in the Detail and the Solution is in All of Us, *Environment and Urbanization* 23, no. 1 (2011): 37, https://doi.org/10.1177/0956247811398588.

Stiglitz, Joseph E. The Price of Inequality: How Today's Divided Society Endangers Our Future, 1st ed (New York: W.W. Norton & Co, 2012)

Stone, Michael E. Housing Policy Debate What is Housing Affordability? The Case for the Residual Income Approach, *Housing Policy Debate* 17 (2010): 37–41, https://doi.org/10.1080/10511482.2006.9521564

Sverdlik,Alice, Ill-health and Poverty: A Literature Review on Health in Informal Settlements, *Environment and Urbanization* 23, no. 1 (2011): 123–55, https://doi.org/10.1177/0956247811398604.

Thünen, J. H. Von. *Der Isolierte Staat (The Isolated State)* (Hamburg: Perthes, 1826). In: https://doi.org/Cited By (since 1996) 13rExport Date 3 February 2012.

Trivelli, P. *Elementos teóricos para el análisis de una nueva política de desarrollo urbano* (Santiago de Chile: Institute of Urban Development Planning, 1981).

Trivelli, Pablo. (2017). Characterisation of areas with potential for densification in pericentral districts of Santiago. Conference at CEDEUS.

Truffello, Ricardo & Rodrigo Hidalgo, Policentrismo en el Área Metropolitana de Santiago de Chile: reestructuración comercial, movilidad y tipificación de subcentros, EURE (Santiago) 41, no. 122 (2015): 49–73, https://doi.org/10.4067/S0250-71612015000100003.

Universidad Alberto Hurtado, *Encuesta Chile Dice* (Hurtado: Universidad Alberto Hurtado, 2017).

Valencia, Marco. The Free Market City. *Diseño Urbano & Paisaje Magazine* 3, no. 7 (2006).

Vekemans, Roger, Jorge Giusti, and Ismael Silva, *Marginalidad, Promoción Popu- lar e Integración Latinoamericana*, ed. DESAL, Ediciones (Buenos Aires, 1970).

Vergara Perucich, F. and Boano, C. Under Scarcity, is Half a House Enough? Reflections on Alejandro Aravena's Pritzker. *Architecture Review* 21, no. 31 (2016): 37–46. En: doi:10.5354/0719-5427.2016.42516

Vergara Perucich, Francisco. Acts of Dissent as Democratising Urbanism: Polit- ical Space in Santiago de Chile. *Urbe. Revista Brasileira de Gestão Urbana* 11 (2019).

Vergara-Perucich, F. and Aguirre- Núñez, C. Inversionistification in Latin America: Problematising the Market (2019). of renting for the Chilean case. Habitat y Sociedad 1(12), pp. 1–20.

Vergara-Perucich, Francisco. *Urban Design Under Neoliberalism: Theorising from* Santiago (Oxon and New York: Routledge, 2019)

Vergara-Perucich, José Francisco , Juan Correa-Parra, y Carlos Aguirre-Nuñez, The Spatial Correlation between the Spread of Covid-19 and Vulnerable Urban Areas in Santiago de Chile, *Critical Housing Analysis* 7, no. 2 (2020): 21–35, https://doi.org/10.13060/23362839.2020.7.2.512.

Vergara-Perucich, Jose Francisco & Camillo Boano, Vida urbana neoliberal: estudio de factores de jerarquización y fragmentación contra el derecho a la ciudad en Chile,

Revista de Direito da Cidade 11, no. 3 (2019): 426–52, https://doi.org/10.12957/rdc.2019.38541.

Vergara-Perucich, José Francisco. Determinantes urbanos del precio de la vivienda en Chile: una exploración estadística, *Urbano* (2021): 40–51.

Vergara-Perucich, José-Francisco ¿Qué tan caro es vivir en las capitales regionales? La necesidad de descentralizar las mediciones sobre el costo de vida en Chile, en *El nuevo orden regional. Construcción Social y Gobernanza del Territorio*, ed. Verónica Fuentes, Egon Montecinos, y Pedro Güell, 1st ed (Valdivia: Universidad Austral de Chile, 2020), 145–58.

Vergara-Perucich, José-Francisco, Juan Correa, and Carlos Aguirre-Núñez, *Atlas de Indicadores Espaciales de Vulnerabilidad Ante El COVID-19 En Chile* (San- tiago: Centro Producción del Espacio, 2020).

Vergara-Perucich,José Francisco & Aguirre-Nuñez, , Housing Prices in Unregulated Markets: Study on Verticalised Dwellings in Santiago de Chile, *Buildings* 10, no. 1 (2019): 6. https://doi.org/10.3390/buildings10010006.

Vergara, Jorge Eduardo, Verticalización. la edificación en altura en la región metropolitana de Santiago (1990–2014), *Revista INVI* 32, no. 90 (2017): 9–49.

Warren, E., and Warren-Tyagi, A. *All Your Worth* (New York: Free Press, 2005).

Winn, Peter, "No Miracle for Us": The Textile Industry in the Pinochet Era, 1973–1998, en *Victims of the Chilean Miracle: Workers and Neoliberalism in the Pinochet Era, 1973–2002*, 2004, https://doi.org/10.1128/AAC.00308-15.

Zizek, Han and Butler in: Pablo (editor) Amadeo, *Sopa de Wuhan. Pensamiento Contemporáneo En Tiempo de Pandemias*, ed. Pablo Amadeo (Santiago: ASPO (Aislamiento Social Preventivo y Obligatorio), n.d.).

Index

Printed in the United States
by Baker & Taylor Publisher Services